Research Issues in Systems Analysis and Design, Databases and Software Development

Keng Siau
University of Nebraska – Lincoln, USA

IGI PUBLISHING

Hershey • New York

Acquisition Editor:	Kristin Klinger
Senior Managing Editor:	Jennifer Neidig
Managing Editor:	Sara Reed
Assistant Managing Editor:	Sharon Berger
Development Editor:	Kristin Roth
Copy Editor:	Shanelle Ramelb
Typesetter:	Sharon Berger and Jamie Snavely
Cover Design:	Lisa Tosheff
Printed at:	Yurchak Printing Inc.

Published in the United States of America by
IGI Publishing (an imprint of IGI Global)
701 E. Chocolate Avenue
Hershey PA 17033
Tel: 717-533-8845
Fax: 717-533-8661
E-mail: cust@igi-pub.com
Web site: http://www.igi-pub.com

and in the United Kingdom by
IGI Publishing (an imprint of IGI Global)
3 Henrietta Street
Covent Garden
London WC2E 8LU
Tel: 44 20 7240 0856
Fax: 44 20 7379 0609
Web site: http://www.eurospanonline.com

Library of Congress Cataloging-in-Publication Data

Research issues in systems analysis and design, databases and software development / Keng Siau, editor.
     p. cm.
  Summary: "This book is designed to provide understanding of the capabilities and features of new ideas and concepts in the information systems development, database, and forthcoming technologies. It provides a representation of top notch research in all areas of systems analysis and design and database"--Provided by publisher.
  Includes bibliographical references and index.
  ISBN 978-1-59904-927-4 (hardcover) -- ISBN 1-59904-928-1 (ebook)
  1. System design. 2. System analysis. 3. Computer software--Development. I. Siau, Keng, 1964-
  QA76.9.S88R465 2007
  003--dc22
                              2006039749

British Cataloguing in Publication Data
A Cataloguing in Publication record for this book is available from the British Library.

Advances in Database Research Series

ISSN: 1537-9299

The Advances in Database Research (ADR) Book Series publishes original research publications on all aspects of database management, systems analysis and design, and software engineering. The primary mission of ADR is to be instrumental in the improvement and development of theory and practice related to information technology and management of information resources. The book series is targeted at both academic researchers and practicing IT professionals.

Order online at www.igi-pub.com or call 717-533-8845 x10 –
Mon-Fri 8:30 am - 5:00 pm (est) or fax 24 hours a day 717-533-8661

Research Issues in Systems Analysis and Design, Databases and Software Development

Table of Contents

Preface

Revolution and evolution are common in the areas of information systems development (ISD) and databases. New concepts such as agile modeling (AM), extreme programming (XP), knowledge management, and organizational memory are stimulating new research ideas among researchers and prompting new applications and software from practitioners. This volume, *Research Issues in Systems Analysis and Design, Databases and Software Development*, is a collection of state-of-the-art research-oriented chapters on information systems development and databases. This volume does not only serve the research purposes of researchers and academicians, but it is also designed to provide technical professionals in the industry with understanding of the capabilities and features of new ideas and concepts in information systems development, databases, and forthcoming technologies.

Keeping with the high standard of previous volumes in the *Advances in Database Research* series, we carefully selected and compiled 10 excellent chapters written by well-known experts in the areas of information systems development and databases. A short description of each chapter is presented below.

Chapter I, "Agile Software Development in Practice," explores agile information practices of information systems development and argues that their history is much longer than what is generally believed today. It takes an interpretive and critical view of the phenomenon. This chapter reports an empirical study on two companies that apply an XP-style development approach throughout the information systems development life cycle.

Chapter II, "Understanding Agile Software, Extreme Programming, and Agile Modeling," discusses the state of research in extreme programming and agile modeling. This chapter also examines research in agile software development. It first presents the details of agility, XP, and AM, including a literature review, followed by an identification of gaps in the literature and a proposal for possible future studies.

Chapter III, "Adaptation of an Agile Information System Development Method," presents the work practice in dealing with the adaptation of an agile information systems development method in the ISD department of one of the leading financial institutes in Europe. This chapter also introduces the idea of method adaptation as an underlying phenomenon concerning how an agile method has been adapted to a project situation in the case organization.

Chapter IV, "Matching Models of Different Abstraction Levels: A Refinement-Equivalence Approach," discusses the reuse of models, which assists in constructing new models on the basis of existing knowledge. It proposes the concept of refinement equivalence and emphasizes its use for the purpose of validating a detailed application model against an abstract domain model in the context of a domain analysis approach called application-based domain modeling.

Chapter V, "On the Use of Object-Role Modeling for Modeling Active Domains," discusses how the object-role modeling (ORM) language and approach can be used for integration, at a deep and formal level, of various domain-modeling representations and viewpoints, with a focus on the modeling of active domains. The chapter argues that ORM is particularly suited for enabling such integration because of its generic conceptual nature; its useful, existing connection with natural language and controlled languages; and its formal rigor.

Chapter VI, "Method Chunks to Federate Development Process," proposes an approach that consists of federating the method chunks built from the different project-specific methods in order to allow each project to share its best practices with the other projects without imposing on all of them a new and unique organization-wide method.

Chapter VII, "Modeling and Analyzing Perspectives to Support Knowledge Management," introduces a generic modeling approach that explicitly represents the perspectives of stakeholders and their evolution traversing a collaborative process. This chapter also describes a Web-based information system that uses the perspective model and the social-network analysis methodology to support knowledge management within collaboration.

Chapter VIII, "Modality of Business Rules," discusses one way to model deontic rules, especially those of a static nature. A formalization based on modal operators is provided, and some challenging semantic issues are examined from both logical and pragmatic perspectives.

Chapter IX, "Lost in Business-Process Model Translations: How a Structured Approach Helps to Identify Conceptual Mismatch," discuses the problem of translating between process modeling languages. It argues that there is conceptual mismatch between modeling languages stemming from various perspectives of the businesses-process management life cycle that must be identified for seamless integration.

Chapter X, "Theories and Models: A Brief Look at Organizational Memory Management," introduces theories and models used in organizational memory. This chapter provides a brief review of the literature on organizational memory management and further presents a basic framework of theories and models, focusing on the technological components and their applications in organizational memory systems.

The 10 chapters in this volume provide a snapshot of the latest research in the areas of information systems modeling, systems development, and databases. This volume is a valuable resource for scholars and practitioners alike.

Professor Keng Siau, PhD, University of Nebraska – Lincoln
E.J. Faulkner Professor of Management Information Systems
Editor-in-Chief, Advances in Database Research Book Series
Editor-in-Chief, Journal of Database Management

Chapter I

Agile Software Development in Practice

Matti Rossi, Helsinki School of Economics, Finland

Hilkka Merisalo-Rantanen, Helsinki School of Economics, Finland

Tuure Tuunanen, The University of Auckland, New Zealand

Abstract

This chapter explores agile information practices of information systems development and argues that their history is much longer than what is generally believed today. We take an interpretive and critical view of the phenomenon. We made an empirical study of two companies that apply an XP-style development approach throughout the information systems development life cycle. The results of our research suggest that XP is a combination of best practices of traditional information systems development methods. It is hindered by its reliance on talented individuals, which makes its large-scale deployment as a general-purpose method difficult. We claim that XP can be useful for small colocated teams of skilled domain experts and implementers who are able to communicate well with the end users. However, these skilled and motivated individuals with high working morale can exhibit high productivity regardless of the methods used if they are not overly constrained by bureaucracy.

Introduction:
From Methodologies to Methods and Agility

Ever since the first major software systems were developed, a chronic "software crisis" has been seen either looming ahead or haunting the community (Brooks, 1975). Solutions have been sought mostly in raising the productivity of programmers, making systems less defective (e.g., process management and development approaches; Boehm, 1988; McConnell, 1996), and developing systems by methods that treat the end users as equals to the designers in the development process (e.g., participatory design, PD; Bjerkenes & Bratteteig, 1995; Grudin, 1991). In this chapter, we first discuss these approaches for organizing information systems development (ISD). This leads us to a discussion of agile software development methods that have emerged as a fresh alternative for the more rigid life-cycle-based approaches in recent years.

Extreme programming (XP) tries to address end-user participation and increased quality of work by emphasizing the use of professional work practices and ethical software development. The waterfall model emerged as a systematic, sequential solution to software development problems (Brooks, 1975; Hirschheim, Klein, & Lyytinen, 2003). The IS product was not delivered until the whole linear sequence had been completed. As projects became larger and more complex, problems like stagnant requirements and badly structured programming started to arise.

Overlapping the phases (Fairley, 1985; Pressman, 2000; Sommerville, 2001) and the introduction of the more incremental spiral model (Boehm, 1988; Iivari, 1990a, 1990b) resolved many of the difficulties mentioned earlier. This model presents the software process as a spiral, where each of the loops can be considered to represent one fundamental development step. Thus, the innermost loop might be concerned with requirements engineering, the next with design, and so on (Sommerville). The spiral model assumes a risk-driven approach to the software development rather than a primarily document-driven (waterfall) or code-driven (prototyping) approach (Boehm). Each cycle incrementally increases the system's degree of definition and simultaneously decreases its degree of risk (Boehm, Egyed, Kwan, Port, & Madachy, 1998).

The iterative models were augmented with more dynamic approaches with less bureaucracy. For example, in incremental development, software is developed in small but usable pieces that can be delivered early on to a customer. Each increment is an operative subset of the final software system and builds on the increments that have already been developed (Pressman, 2000).

Parallel to ISD organization changes, the design craft itself has been evolving. It has been argued (McKeen, Guimaraes, & Wetherbe, 1994, pp. 427-428) that user participation improves the quality of the system in several ways such as "providing a more accurate and complete assessment of user information requirements ... providing expertise about the organization the system is to support ... avoiding development of unacceptable or unimportant features, and improving user understanding of the system ..." Nevertheless, there was no common definition of how users should be involved (Carmel, Whitaker, & George, 1993). To solve this problem, many approaches arose, most notably PD (Bjerkenes & Bratteteig, 1995) and joint application development (JAD; Clemont & Besselaar, 1993). While taking a different view of end users' role, both stress the involvement of users in the development process and design decisions. New methods and tools to help communication among IS designers and users are continuously developed (e.g., Liu, Pu, & Ruiz, 2004; Shoval & Kabeli, 2001). One of the key arguments of this discussion has been how to reconnect the designer and user again (Grudin, 1991).

The last aspect that agile approaches, and especially XP, raise is the empower-ment and productivity increase of developers. Traditionally these have been sought by raising the abstraction level of the software development tools (e.g., through high-level languages and CASE). However, programmers have often seen these more as an obstacle. One suggested solution is the employment of work practices that let the most talented developers unleash their power (e.g., surgical teams [Brooks, 1975] and pair programming, which, according to Williams & Kessler, 2002, dates back to Brooks in the 1950s).

To conclude, XP seeks to solve many of the problems of traditional software development by combining the best practices from the past research and practice of ISD. First, XP aims at employing participatory design by really engaging the business or end users into the IS development process. Second, XP seeks to add flexibility to the development process and to organize the work into small packages with clear deliverables. Finally, XP tries to squeeze maximal productivity out of the developers by using concepts such as pair programming.

In this study we explore the agile software development approaches as they appear in practical context. We argue that XP can be described as a way of working that codifies old practices rather than creates new ones. However, we argue that XP may add some value into the development-process discussion as it connects prototyping and end-user-oriented development in a way that could deliver systems that are a better match for the end-user needs. We explore these arguments in the following by first looking at XP and its roots in agile methods in the second section. This is followed by the description of the methodology and the presentation of the case studies. Thereafter, we discuss the findings from the case companies. In the final section we draw conclusions based on the cases and point out future research challenges.

Agile Methods and Extreme Programming

There are about a dozen software development approaches that are classified or regarded as agile—XP being the most popular of them. Common to all agile methods is the emphasis on the output of the software development process, working software, and maximizing its value for the customer. Agile methods are mostly used when developing tailored software in house. Agile software development methods can be defined as using human- and communication-oriented rules in conjunction with light, but sufficient, rules of project procedures and behavior (Cockburn, 2002). These four rules are individuals and human interactions over processes and tools, working software over comprehensive documentation, customer collaboration over contract negotiation, and responding to change over following a plan (*Agile Manifesto*, 2003). The emphasis on communication and programmers' morale is common to all agile methods. In accordance with Conrad (2000), agile methods focus on people as the primary drivers of development success. In the following, we focus on key principles of one agile method, extreme programming, first introduced by Kent Beck (1999). For a more detailed overview of agile methods in general, see, for instance, Abrahamsson (2003) and Abrahamsson, Warsta, Siponen, and Ronkainen (2003).

According to Beck (1999), "XP is a lightweight methodology for small-to-medium-sized teams developing software in the face of vague or rapidly changing requirements" (p. xv), and "XP is a lightweight, efficient, low-risk, flexible, predictable, scientific, and fun way to develop software" (p. xvii). In turn, Abrahamsson et al. (2003, p. 245) have defined XP as a "collection of well-known software engineering practices ... . The novelty of XP is based on the way the individual practices are collected and lined up to function with each other."

XP addresses risk and value of software at all levels of the development process. According to Beck (1999), customers (or managers) can pick three out of four control variables (these are cost, time, quality, and scope) and the development team decides on the fourth. Technical people are responsible for work estimates, technical consequences of business decisions, the development process, and detailed scheduling within a release. Team size should be in maximum about 12 designers and the software not excessively complex (Beck).

The project management strategy of XP maximizes the value of the project by providing accurate and frequent feedback about progress, many opportunities to dramatically change the requirements, a smaller initial investment, and the opportunity to go faster. In XP, cost, time, and the quality of a component are regarded as fixed control variables decided by customers and managers. Within these limits the development team focuses on the variable development scope, that is, on the functionality of the parts. The programming strategy of XP is to keep the code easy to modify (Beck, 1999).

The 12 principles or rules of the XP methodology are planning, small releases, metaphor, simple design, testing, refactoring, pair programming, collective ownership, continuous integration, having the customer on site, coding standards, and a 40-hour week (Beck, 1999).

The principal values are communication, simplicity, feedback, and courage. The effect of stressing testing, pairing, and estimating in the development process is that programmers, customers, and managers have to communicate. Simplicity means doing the simplest thing that could possibly work and adding complexity later if it is really needed. Feedback works on different time scales: minutes, days, weeks, and months. Courage is needed to change the basic architecture or to code a piece of software again from scratch. Basic principles for decision making are derived from these values: rapid feedback, simplicity, incremental change, embracing change, and quality work (Beck, 1999).

In Figure 1 we show a slightly modified XP approach called the agile development approach (ADA), which was identified in our Case Company 1. In this model, we have depicted the agile information-development tasks in phases and the outputs of each phase. An XP project begins with a task called an architectural spike. The outcome of this task is a system metaphor, that is, the

Figure 1. Agile development approach (ADA)

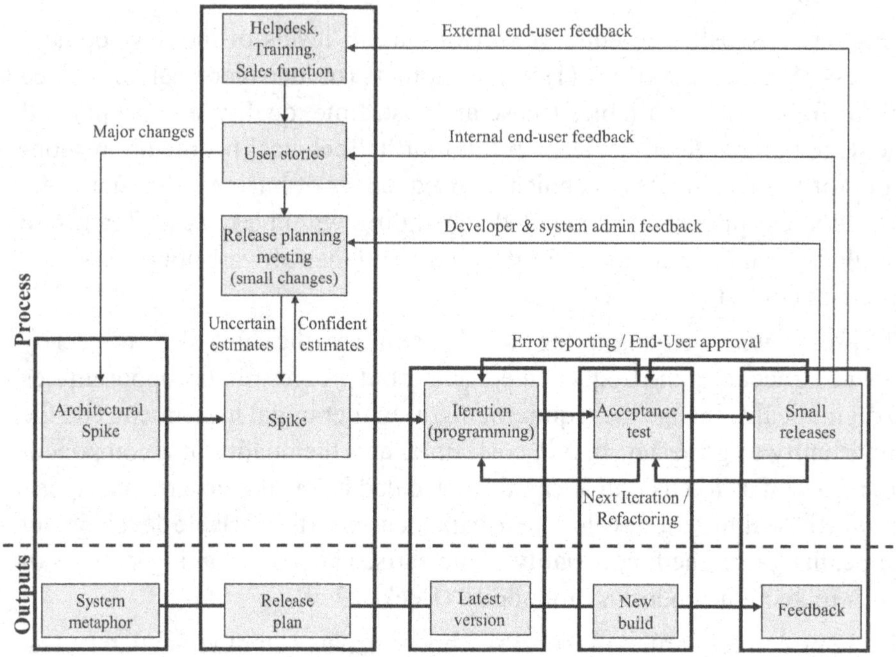

infrastructure, standards, and working habits of the XP project. Beck (1999) says that a metaphor is a simple story of how the whole system works, for instance, an outsourcing contract or software architecture. It helps everyone in the project to understand the basic elements and their relationships, and it is easy to communicate and elaborate on. Both business and technical people participate actively in the definition of the system metaphor.

User stories are task descriptions, also called user requirements or user needs, and possibly also descriptions of expected benefits. End users write user stories in plain text using their own terminology. Developers estimate the ideal development time of the story, which can vary from 1 to 3 weeks. User stories must be combined or broken down if these limits are not reached. A spike solution is programmed if it is needed to make the estimates more accurate. A release plan lays out the overall project. It specifies which user stories are going to be implemented for each release and when each release will be finished. If the velocity of the development changes dramatically, a new release-planning meeting should be held to reestimate and renegotiate the release plan.

The iteration task of an XP project produces a new version or release of the program in progress for acceptance tests. A program or piece of code is integrated into the system either after it has passed all the unit tests or after some smaller part of the planned functionality has been finished. Each developer must integrate and release his or her code every few hours or at least once a day. This kind of continuous integration avoids and detects compatibility problems early, and everyone always works with the latest version of the system. This approach also avoids many of the problems of too rigorous and formal approaches by stressing increments and iteration over rigor and waterfall development (Joosten & Purao, 2002). In XP, coding is done in pairs on one workstation, and pairs are changed continuously. The code should be collectively owned and each programmer is allowed to change the code, with changes done by one programmer at a time. The code is refactored continuously to improve its quality and to make it as simple as possible without making any changes to its functionality or its features. However, pair programming was not used in either of our two cases.

Acceptance tests are run on the latest version of the system to ensure the functionality of the system. End users are responsible for acceptance tests, and they specify the test scenarios based on user stories. They also review the test scores and prioritize the corrections needed.

Finally, after the end user or customer has approved a small unit of functionality, it is released into the customer's environment. Small, frequent releases give a possibility to get feedback from the users early on and to make changes into the release plan if necessary. We have complemented our ADA model in Figure 1 with a help desk, training, and sales function so that indirect user requests can also be easily gathered. We have also added the feedback arrows to emphasize the possible effects of the end-user feedback. Very often the feedback may be the driver for modified or new user stories that is also common with more traditional ways of development (Boehm, 1988). Similarly, feedback may result in changes to the release plan. Furthermore, it is possible, although exceptional, that the architectural spike of the project is revised.

Methodology of This Study

Some recent research has focused on the planned and systematic adoption of XP in different contexts (Abrahamsson, 2003; Abrahamsson, Salo, Ronkainen, & Warsta, 2002; Back, Milovanov, Pores, & Preoteasa, 2002; Elssamadisy & Schalliol, 2002; Reifer, 2002; Salo & Abrahamsson, 2004). However, we were not able to find many studies focusing on the natural evolution of ISD practices toward more agile approaches. Notable exceptions are Aoyama (1998), Murru, Deias, and Mugheddu (2003), and Vanhanen, Jartti, and Kähkönen (2003), who describe the institutionalization of agile practices in organizations. Hence, we wanted to study further why and how XP is adopted and used in everyday software production. Furthermore, we were interested in seeing whether the method was intentionally selected or if it had gradually evolved based on the methods used before.

We decided to take an interpretive but, at the same time, critical approach (Myers, 1997). We followed the guidelines of Klein and Myers (1999) and adopted qualitative research as a means of trying to understand this complex and fast-moving IS research topic. We turned to the case-study approach that Wynn (2001) has advocated as the most appropriate qualitative method in studying social processes and trying to understand users at the local level. In the case descriptions we adopted the principles of interpretive case studies presented by Walsham (1995) in contrast to the positivist approach to case studies. These principles are reporting details of the selected research sites, the reasons why these sites were chosen, the number of people interviewed, the interviewees' hierarchical or professional position, secondary sources of data, the data-gathering period, how field interviews and other data were recorded, the description of the analysis process, and finally, how the iterative process between field data and theory took place and evolved over time.

The case companies were selected in two phases. We began by actively seeking companies that use agile development practices and tried to identify potential candidates for our study. We did this by gathering information from other researchers of agile methods in Finland and discussing with ISD personnel in several potential case companies concerning the development methods in use. Thereafter, two companies employing agile practices were selected. The case companies were intentionally selected from different industries: a manufacturing company vs. a software-developer consultancy. The companies also differed greatly in their reasons for selecting this kind of approach to IS development and the drivers behind its adaptation. The first case firm had gradually evolved their own method or way of working, whereas the second case company had made a more or less deliberate decision to employ agile development practices.

In each company we focused on one major system or software central to the company. The systems were in the maintenance phase and under continuous renewal after several years of development. We conducted semistructured theme interviews with two IT managers, a business-development manager, and a senior consultant. We also received written documents as well as other complementary information on the IS development processes. Later on, the data were complemented by telephone discussions and e-mails. The data collection was conducted during spring 2003. The interviews were tape-recorded and transcribed, and later validated with the interviewees. The interviewees also verified and accepted the final version of the case descriptions. The questions of the semistructured interviews are available from the authors on request.

The data analysis was done by comparing the interview data to the general ISD process literature with focus on agile methods. More specifically, we sought to understand how companies applied agile practices. This iterative process is reported in the following sections in more detail.

Cases

In this section we present two cases of employing agile practices in software production: a factory system and a communications application portfolio. Each case begins with a short description of the company and the system. Thereafter, the drivers of the development of the case system and the used methods are described. Then the software development process is delineated following Figure 1. Finally, some insights of the interviewees concerning the contemporary software process and its future are represented. The development organization, users, and tools are described in more detail in Appendix A for the first case, and in Appendix B for the second case. The findings of the cases are presented and discussed in the next section.

Case 1: Factory System

Case Company

The case organization is a processing division of an international group in the industry. The information technology function of the division is located in Finland on the premises of a factory. A separate information technology unit was merged with the business-development unit, which is also in charge of the development of production planning, logistics, and procurement functions. The information technology function employs 12 people divided into three teams. The first team of six employees, the information systems development team, is in charge of the factory system. The second team is responsible for the management information systems, statistics, and packaged software used in these activities, and the third team is responsible for hardware, networks, and packaged software except those used in management information systems and statistics.

Factory System

The studied system, later called the factory system, has been developed in house. The first small application was developed in 1986. The factory system consists of three main parts: a sales system, a mills or production system, and a business reporting system, with the main applications being sales, production (i.e., factory), maintenance, purchasing, and statistics. Practically all employees are end users of the factory system. A software package for online analytical processing (OLAP), multidimensional analysis, and reporting is integrated into the factory system.

The factory system enables the factories and sales network to monitor production and deliveries in real time. The logistics services simplify the ordering process. The material requirements of the customers are monitored and predicted in real time. This means that storage needs are reduced and customers are assured of getting their material at all times. Delivery reliability is based on cyclical production and standardized, uniform grades delivered in a standard pack size. Customers can consult the case organization on matters relating to the choice of material and the design of the product. The whole production is conducted according to the customer needs, which is still exceptional in this field. The development organization, users, and tools are described in more detail in Appendix A.

Drivers

The motto of the factory-system development is: "To know the business, to stay in it and to be the best in the business. The system is tailored to the business, not to the company." The key driver of the development work is the continuously changing and increasingly complex business environment. In addition, the development perspective is clearly bottom-up as user needs drive the continuous development of the system.

The key factor for the success of the development work is the domain knowledge of the team members. Each of them has a long experience in other companies and in different jobs, as well as a deep understanding of the business. Expertise with the tools used is not seen as equally important. A person must be extroverted, speak the language of the users, be actively in contact with users, and be easy to approach. Responsibility, initiative, and desire as well as the ability to inform others are also seen as important characteristics.

Architectural Spike: Methods and Standards

The current development tools (see Appendix A) were selected around 1986 when a decision was made to move over from minicomputers to microcomputers and client-server architecture. One designer was responsible for selecting new tools for this critical 24/7 system. The first small application with the current tools was developed and taken into production in 1986, with an expert from the vendor participating in the development and training the first users who gradually took over the development work.

The original factory system with current tools was developed during 1986 to 1990 as a project following the waterfall development model. Because of the long and slow development phase, the system had gone out of date by the time it was finished. The contemporary working method was introduced in 1990 when the development of the current factory system began. Since then, the method has evolved and become more and more streamlined.

The working methods and standards are discussed and, if necessary, changed by developers weekly. Coding rules especially are strictly standardized from the number of empty rows between the program parts to the starting position of a certain row type. This enables common code and makes the reading of the code quicker.

Nowadays the method is used throughout the development life cycle. No other development methods are used and no project work is done either because of their slowness and inflexibility.

User Stories, Release Planning, and Spikes

Requirements are actively collected from many sources. Business goals, objectives, and structures are received from management as policy statements. Requests for technical changes are usually raised by the system administration. User requirements are actively collected, and daily communication between developers and end users, both official and spontaneous, on the factory premises is extremely valuable. Now, in the production phase, user requirements are received from users through a system help desk or as feedback given directly and informally to the developers. All the feedback is registered. The request for a change or a new requirement may also stem from the lack of a function in the system or the inability of the system to serve a function. Also, the need to reduce staff from a function or the shortage of employees in a function may give an initiative to system improvement.

Developers go through user feedback daily to find and fix errors. These corrections, and small improvements, are usually installed immediately. Requirements and feedback are also gathered, and the management looks through this list regularly and decides on future development. A certain part of the system will be renewed if there are plenty of negative comments concerning it. The number of users is often used as a decision criterion. Costs or the time schedule have less value in decision making. For a major renewal, both a short-term (1 month) work plan and a long-term (6 months) iteration plan are drafted. All the designers work only part time with major renewals. A spike is programmed if necessary to ensure the feasibility of the planned solution.

Iteration, Refactoring, Testing, Releases, and Pair Programming

A new program is initiated first by coding a program skeleton with only a few basic functions. This semifinished program is installed into the production environment to make sure that it will meet the basic requirements. At the beginning, the developer uses the skeleton parallel to the old one, if it exists, and collects experience of its operation. The skeleton is continuously changed and extended; that is, it will never be finished. Its development will be stopped for a while once a desired service level is attained. The developers know by experience and domain knowledge when this level is reached.

A developer may make small and simple changes and error corrections directly into the production environment independently. If changes are needed in other parts of the system or in the database, they are made first in a test environment, a copy of the production environment, on the developer's own microcomputer. When all the changes have been finished, they are installed into the production environment. It is the responsibility of the developer to make sure that any changes made by other developers directly into the production environment in the meantime stay in use and perform as expected. All changes are registered into a change log file.

A developer tests his or her own code. In addition, the two team members responsible for training and the help desk will test larger changes before taking them into production.

All changes must be made and installed directly into the production environment of the system incrementally because the factory operation depends entirely on this system and operates 24/7 shifts. A development phase will take from hours to a few days or sometimes a few weeks, but never months.

Pair programming is not used in the case company and the developers do not share work premises. Error and problem solving, and spikes are done in pairs, if necessary.

Documentation

Documentation has been reduced to an absolute minimum and only the key (useful) documents are drafted and maintained. Working methods, standards, and coding rules are documented as an instruction sheet. Developers are responsible for database diagrams, entity-relationship diagrams, and change log files. Program code is the most important document for the developers. Trainers are responsible for the user's manual or system help. This manual is an Acrobat PDF (portable document format) file consisting of instructions for managing special cases or functions, each of them being at most one sheet long.

Future

This development method and the efficient development tools in use meet the needs of the constantly changing business best. They enable quick and continuous modification and evolution of the system. The development work is done in small incremental pieces so nothing or very little is wasted and hardly anything useless is done.

The way of working is also inexpensive. The total information technology expenses, according to the company, are far below the average in the sector because the system is built in house and practically no software or license fees are paid or external work force is needed.

The information systems development will be continued in this well-working manner. Every other method would require more bureaucracy and strict responsibilities. The present employees have deep knowledge of the business domain, enabling them to work independently and with low hierarchy. In addition, the employees argued that they would probably suffer from lack of motivation if some other working practices were used.

Case 2: Communications Application Portfolio

Case Company

The case organization is a corporate communications agency that was founded in 1986. It is a part of an international network of advertising, marketing, communications, and interactive agencies. The agency provides services ranging from communications research, strategic planning, crisis communications, public relations (PR), public affairs, corporate communications, and investor relations to print publications. The customer companies come from diverse industries as well as from various governmental and municipal organizations. The agency employs around 50 professionals.

An increasing pressure to move and extend traditional communications to incorporate a digital presence and form led to the establishing of a digital communications unit in May 2001. The business strategy of the unit is to support and extend traditional communications and PR activities offered by the agency. The unit employs 12 people divided into three teams of four: consulting, graphics, and technology. The technology team is in charge of the application portfolio, user-interface design, Web site production, updating and maintenance services, and application service provision (ASP), which is the most common way to use the customer software.

Communications Application Portfolio

The communications application portfolio is a software tool kit developed in house. The development work began in May 2001 and took about 18 months. Now the portfolio is in the maintenance phase and is used in customer software development. The portfolio consists of four applications: extranet, content management, monitoring application, and crisis communications, which are described next.

The extranet is a low-risk and low-return application used only in project management for coordinating and facilitating communications between the company and its customers. The extranet has a supporting role, but it is essential for the success of customer projects. It needs only a little nonnative code and customization.

The content-management application provides a Web interface for the creation, management, and maintenance of different forms of content. It is a communications solution and often has a critical function in customers' activities. Some customization is required with each individual implementation.

The monitoring application is primarily used as a business-intelligence tool to collect and store digital information. It acts as a search engine querying specified keywords from predefined Web sites and newsgroups. Matches are stored in its database and a summary of the findings is presented to the user. Customization is always required.

The crisis-communications application is targeted at an extremely narrow, niche audience. It is a collection of Web services to help administer and manage a crisis from a communications point of view. It includes, for example, a digital version of the customer's crisis manual, holding statements, decision-support trees, press contact information, and crisis scenario planning information. It holds the greatest future potential as currently the case company is the only provider of such an application. The user interface is standardized, but all the underlying services are customized.

The development organization, users, and tools are described in more detail in Appendix B.

Drivers

The initiative for the communications application portfolio development came from the need to reuse and standardize the existing modules and programs. The main focus of the development is the consolidation of separate key software modules and individual applications, which have arisen from either individual development projects or as part of customer projects. The gains sought are a common platform reducing development and maintenance costs and time, differentiation from competitors, and flexible pricing mechanisms.

The key factor for the success of the development work is the experience and expertise of the team members with the tools and technology used. Expertise with the business is not equally important. Also, the team spirit—the programmers being familiar with each other before starting to work together—is essential.

Architectural Spike: Methods and Standards

The technology philosophy is derived from the unit's strategy. When the unit was founded, the key question regarding technology was not which one to use and how to implement it, but rather how to acquire skilled people that fit the organization. A collective decision among the original employees, two consultants, and a programmer was made on the technologies and platforms (see Appendix B). The main reason for selecting these tools was that the employees were familiar with them beforehand. No rigorous selection process was conducted. The methods, standards, and working habits employed in software development evolved along with the portfolio development without any conscious decision.

The selected way of working is used in the product development and maintenance phases as well as in customer implementations. The original communications application portfolio was not developed as a project. Customer implementations are always carried out as projects with a nominated project organization using a waterfall model with clear sequential phases.

No written standards or rules concerning the process or the code exist. Rather, all developers have the freedom to decide about their own work. However, unwritten rules exist, and these have been adopted easily because the programmers have a common background.

User Stories, Release Planning, and Spikes

A requirement or need, either functional or technical, can come from employees or customers (50/50). Each customer has a nominated contact person who receives the feedback. New technical requirements often evolve from technical development of the tools or from system administration. A quick situation analysis is performed to assess the feasibility of the requirement.

A programmer independently makes small improvements and error corrections. More complex and far-reaching decisions, like new functions, interfaces, or integration needs, are dealt with collectively. Also, decisions concerning the implementation of these features as development investments or customer projects are made together based usually on the number of potential customers and end users. Spikes are seldom programmed.

Iteration, Refactoring, Testing, Releases, and Pair Programming

The development process has four distinct phases: the initial, planning, development, and maintenance phases. The decision in the initial phase to proceed with a requirements specification leads to the planning phase. In the planning phase, the first project plan either leads to a document outlining the technical specifications of the new product or to the rejection of the project. For approved projects, a detailed second project plan for the development phase is drafted. The primary decision is the choice between implementing it as a customer project or as a separate product-development project. After the development phase, a maintenance phase outlines the 12-month development road map. The phase continues with iterations and feedback from customers.

Each programmer codes and tests his or her own component or piece of program. Thereafter, internal users representing the customer test the system. External customers test the production version of the system before it is launched into production use. Iterations are continued until the application or customer software is acceptable. The code is also continuously refactored and thoroughly documented.

Pair programming is not used systematically in the case company, although the programmers are sitting at the same table. Error and problem solving is done in pairs if necessary.

Programmers update changes into the ASP customer software gradually. For customers using the software on their own computers, changes are occasionally distributed as an update package.

Documentation

The programmers draft all documentation concerning the application portfolio. Some technical documents are drafted in the planning phase and some systems or integration descriptions in the development phase. Program code is the most important document for the programmers, and it must be documented in detail. The program size should be about 2,000 to 4,000 LOC. No change or problem log is drafted or updated. However, a document or version management system and a component library help to manage physical changes.

The training material and online help for end users are drafted collectively. An implementation manual for main users of the applications is also drafted jointly.

Future

The communications application portfolio is sufficient for contemporary business needs. Significant technological changes might cause the renewal of the portfolio. The personnel may be regarded as permanent in the team.

The total information technology expenses of the team have not been monitored, but salaries form the main part. IT investments are probably rather low because of the use of mostly free open-source software. No software or license fees are paid and no external personnel are used.

The information systems development will continue accordingly at least as long as the existing personnel are employed. Every other method would require more bureaucracy and strict responsibilities, and more personnel would be needed. The employees claimed that they would probably face a downturn in work motivation if some other working practices were used.

Findings and Discussion

In this study we compared the ISD processes of two case companies and their application of extreme programming. We chose one traditional industrial case and one that could be classified as a new-economy case. Interestingly, in the more traditional case, the tools and techniques of XP had been employed for over 10 years and in quite a systematic fashion, though the company had never made a deliberate decision to use XP. In the newer company, the XP process had more or less emerged as a novel way of solving time and budget constraints. The developers were aware of XP practices, but did not choose to engage in it "by the book." This company, with a younger developer staff, had seen agile practices as a natural way of doing things as they did not see the value of more bureaucratic methods. A cross-comparison of the two cases can be found in Table 1.

Findings from the Cases

Many essential features of XP can be found in the working methods of the case organizations as listed in Table 2. The table first lists extreme-programming features slightly adopting the principles and values of XP according to Beck (1999). For each XP feature we identify whether it is used in one of the cases. Furthermore, we identify references from vintage ISD literature to support our claim that these techniques have been in use for a long time.

As can be observed in Tables 1 and 2, both case companies apply XP techniques extensively except for pair programming. In Case 1, XP techniques were used systematically throughout the development life cycle. The method is a result of systematic evolution from stricter methodological practices, which were found to be too restricting and slow. No other development methods were used in Case 1. Project work was also perceived as too slow and inflexible, and nowadays development work is not managed as projects. In Case 2, the programmers utilized the application portfolio in customer projects, so in this aspect the method resembles end-user programming. The key end users, however, are the customers, and customer implementations follow the waterfall model and are organized as projects.

Table 1. Cross-comparison of the cases

Topic	Case 1	Case 2
Case company	• A manufacturing division of an international group, founded about 30 years ago • The operational systems team near users both organizationally and physically	• A PR agency belonging to an international network of agencies, founded in 1986 • The technology team near technology and other team members
System	• The operational system called as the factory system is made in-house • Strategic and critical, 24 hours a day 7 days a week	• The application portfolio is developed in-house • Strategic, not critical
Change	• Continuous and rapid, internal and external in business, system, process, working habits, standards, ownership	• Stable, technology
Driver	• Business driven, not only customer driven approach • Bottom up=user driven, not only management driven approach	• Business driven approach • Technology as enabler of new business possibilities
Methods	• XP, evolutionary prototyping • No other ISD methods • No project work	• XP, waterfall, end-user programming • Customer implementations as projects
Users	• 500 internal end-users	• 4 internal users: 3 internal programmers and a consultant • 300-350 external end-users
Team	• 6 persons, experienced both in business and in technology and methods • Specific roles and responsibilities	• 4 persons, experienced in technology • No specific roles and responsibilities, but one specialist for each application
Requirements	• Business, users, system administration	• Business, technology, customers
Decision making	• Individual developers daily and independently on errors and small changes • Managers (3 persons) together on larger development needs	• Individual programmers daily and independently on errors and small changes, no clear responsibilities • Manager and/or consultant consulted on larger needs and on decisions on customer or development project
Process	• Iterative short cycle process like XP • Resembles XP, but was started in 1990 • Resembles also evolutionary prototyping • No pair programming • Like 1960s – 1970s	• Iterative short cycle process like XP • Resembles XP in some parts, but more like end-user programming or streamlined waterfall • No pair programming

Table 2. Findings from the cases

Extreme Programming Features	Case 1	Case 2	Related ISD Literature
End-user participation	+	-	User centered design (Ehn, 1988; Andersen et al., 1990; Grudin, 1991; Greenbaum & Kyng, 1991; Clemont & Besselaar, 1993; McKeen et al., 1994; Smart & Whiting, 2001)
Standardization of development tools	+	+	Design to tools (McConnell, 1996)
Standardization of development methods	+	-	Any method approach (Paulk, Curtis, Chrissis, & Weber, 1993)
Clear decision rules, roles and responsibilities	+	-	Professional work practices (Andersen et al., 1990)
Distinct phases	+	+	Phased ISD (Hirschheim et al., 2003; Sommerville, 2001)
Iterative development and implementation in short cycles	+	+	Incremental prototyping (Boehm, 1988; Luqi & Zyda, 1990; Iivari, 1990a; Boehm et al., 1998)
Testing	+	+	(Evans, 1984)
Documentation in code	+	+	Literate programming (Knuth, 1984)
Commonly owned program code	+	+	No secret code (Shabe, Peck, & Hickey, 1977)
Specialization on a specific application or on the database or system structure	+	-	Design to tools (McConnell, 1996)
Pair programming	-	-	Fred Brooks in the 1950s (Williams & Kessler, 2002)
Programmer morale	+	+	(Brooks, 1975)
Continuous feedback	+	+	(Boehm, 1988)
Project work	-	+	(Paulk et al., 1993)
Other methods used (waterfall)	-	+	(Brooks, 1975; Hirschheim et al., 2003; Sommerville, 2001)

Way of Working

In the first case, the way of working was adopted as early as 1990, and it has evolved and streamlined gradually and systematically. There is a great resemblance between XP and the development method used in the 1960s and 1970s, when systems were tailored for each organization's own use by IT personnel of its own. At those times, like in the case organization, the basis for future development was both the requirements of business and user needs. Requirements of the management were not a separate matter but they were satisfied through the requirements of the business. In the second case, the information systems development was a new activity in the organization, and the tools and the way of working were introduced and implemented at the beginning.

In both companies, the developers liked this way of working, and the internal and external customers were also satisfied with the results. However, it should be noted that both companies exhibit a key problem of all radically new methods: They are quite person dependent. In the first case we found that XP works best with experienced developers who know their domain and their development tools. The developers were also colocated with the key end users. In the second case, with less experienced developers, we found that the XP development model had more or less emerged instead of having been a planned approach. XP in this latter fashion closely resembles the capability maturity model's (CMM) Level 0, that is, chaos.

Development Organization and Personnel

In Case 1, the information technology unit is part of the business development, and this is crucial for the success of the way of working. On the other hand, the team's physical proximity to the users helps to maintain the knowledge of the business and of user needs, and reduces the dependency on individual developers.

In the first case, the domain knowledge of the team members as well as their excellent communication skills was found extremely important. Without these kinds of persons, the chosen approach would probably have little possibilities to succeed. This was clear also in the second case, where the expertise of the team members with the tools and technology used as well as their own community were extremely important to enable this way of working. The development method was highly dependent on individual programmers, but therefore it suited perfectly the organizational culture of the firm. This finding is consistent with those about the so-called "Internet speed" development (Baskerville, Ramesh, Levine, Pries-Heje, & Slaughter, 2003).

Continuous feedback, both official and unofficial, was one of the key factors of success. In Case 1, very little feedback on the general success of the system is received from current users. Generally, positive feedback is received from users who have left the organization or from newcomers who have the possibility to compare this system with others. There is no change resistance, and users propose changes and improvements to the system actively. They also understand that everything is not reasonable to fulfill, and this fact keeps the method working.

The tools employed facilitated the use of XP in both cases. They supported the fast delivery and easy modification of prototypes. This closely resembles the ideas put forth by early advocates of incremental prototyping (Luqi & Zyda, 1990) and user-centered design (Ehn, 1988), and furthermore design to tools (McConnell, 1996).

Comparison with Other Cases and Agile Methods in General

In this chapter, we took a different approach from other recent case studies of XP (Abrahamsson, 2003; Abrahamsson et al., 2002; Back et al., 2002; Elssamadisy & Schalliol, 2002; Reifer, 2002; Salo & Abrahamsson, 2004), which concentrated on the planned and systematic adoption of XP in laboratory cases or in pilot projects. We selected cases in which the methods had evolved organically into an agile way of working although it was not intentionally and consciously selected as a method. Aoyama (1998) reports evolution and experiences in a telecommunications software family development over a time period of 10 years, very similar to our first case. Likewise, Vanhanen et al. (2003) report the evolution of agile practices in a Finnish telecom industry in three projects, one of which has a life span of over 15 years, again very similar to our first case. In all three projects, agile practices were used (evolved or intentionally adopted) because they represented a natural and useful way to develop software. The authors found that the longest running project applied most widely and systematically agile practices, also similar to our findings.

Opinions differ significantly on the relationship between traditional and agile methods. Some researchers argue that agile methods present an alternative to process-centered approaches (Baskerville et al., 2003; Boehm, 2002; Murru et al., 2003) while others see agile and process-centered methods as complementary (Boehm & Turner, 2003; Paulk, 2001). A third group of researchers see agile processes as a step further in software process improvement as regarded from the CMMI point of view (Kähkönen & Abrahamsson, 2004; Turner, 2002). Increasingly both researchers and practitioners see agile and traditional plan-driven methods as complementary so that different development situations are best supported by different development methods (Boehm & Turner, 2003; Henderson-Sellers & Serour, 2005; Howard, 2003; Känsälä, 2004). Boehm and Turner propose a multidimensional model for selecting the appropriate software development method according to the type of the project. Henderson-Sellers and Serour propose a method engineering approach for assembling agile methods.

To sum up, there are about a dozen software development approaches that are classified or regarded as agile. Common to all agile methods is the emphasis on the output of the software development process, working software, and maximizing its value for the customer. All agile methods, including XP, have their strengths and weaknesses, and different methods are suitable for different business and software development situations. The field is continuously developed further by academics (Nawrocki, Jasinski, Walter, & Wojciechowski, 2002; Visconti & Cook, 2004). Agile methods, like all software development methods, are also continuously evolving through adaptation by practitioners in daily use (Wynekoop & Russo, 1995). The two cases of this research illustrate how practitioners adapt and apply methods. The research provides reasons why practitioners turn to agile methods. It also indicates that the method selection discussion should not be limited to which method is better than the other but instead the discussion should focus on the drivers, constraints, and enablers that affect the selection of the method.

Conclusion

In this study we used a qualitative case-study approach as recommended by Klein and Myers (1999) and Wynn (2001) for studying social processes of agile software development and trying to understand users at the local level. In the case analysis, we adapted the principles of interpretive case studies presented by Walsham (1995). We found support for our claim that XP is more of a new bag of old tricks than a totally new way of doing things. It formalizes several habits that appear naturally in a setting like our first case: close customer involvement, short release cycles, cyclical development, and fast response to change requests. In other words, it combines the best practices of the Scandinavian approach (Bjerkenes & Bratteteig, 1995; Grudin, 1991) in general and user-centered design in particular into a package that is both acceptable and applicable for developers. The so-called Scandinavian approach to information systems development has been advocating user centeredness and professional work practices since the mid '80s, and its roots can be traced back to the origins of object-oriented development (Dahl & Nygaard, 1966). However, it seems that these ideas are easier to accept when they come from within the software development community and have a name that connects them with heroic programming efforts.

It is somewhat disturbing that these practices rely heavily on people and seem to be at times an excuse for not using more refined approaches. We maintain that XP can be useful for small teams of domain experts who are able to communicate well with customers and are very good designers and implementers. One could argue that XP canonizes, and to a certain degree formalizes, the good practices used by these exceptional individuals and teams, which is fine. However, these people can exhibit high productivity in almost any development setting that is not overly constrained by bureaucracy. The real test of XP is, then, whether mere mortals or "normal" developers can employ it as successfully.

In the future, we would like to see how XP can be used in larger scale settings with external customers, either consumers or users in other units within the same company, possibly located in other countries. These would put XP in test with more complex requirements-gathering and -elicitation phases and maintenance of systems through release versions. It would also be interesting to study if XP or some other agile method would be easy enough to be adopted in more traditionally organized IS departments. XP might also be a useful method for organizations with only a few IS specialists in managing their ISD projects with external consultants and vendors.

Acknowledgment

This research was supported in part by the Academy of Finland (Project 674917), the Jenny and Antti Wihuri Foundation, and the Foundation for Economic Education. We wish to thank the contact persons and interviewees in the case companies for their cooperation. We also thank the anonymous referees for their valuable comments.

References

Abrahamsson, P. (2003, September). Extreme programming: First results from a controlled case study. In *Proceedings of the Euromicro 2003*, Antalya, Turkey.

Abrahamsson, P., Salo, O., Ronkainen, J., & Warsta, J. (2002). *Agile software development methods: Review and analysis* (No. 478). Espoo, Finland: Technical Research Centre of Finland, VTT Publications.

Abrahamsson, P., Warsta, J., Siponen, M. T., & Ronkainen, J. (2003, May). New directions on agile methods: A comparative analysis. In *Proceedings of the 25th International Conference on Software Engineering*, Portland, OR.

Agile manifesto. (2003, April 24). Retrieved from http://www.agilealliance.org/

Andersen, N. E., Kensing, F., Lundin, J., Mathiassen, L., Munk-Madsen, A., Rasbech, M., et al. (1990). *Professional systems development: Experience, ideas and action*. Hemel Hampstead: Prentice Hall.

Aoyama, M. (1998, April). Agile software process and its experience. In *Proceedings of the International Conference on Software Engineering (ICSE 1998)*, Kyoto, Japan.

Back, R. J., Milovanov, L., Pores, I., & Preoteasa, V. (2002, May). XP as a framework for practical software engineering experiments. In *Proceedings of the Third International Conference on Extreme Programming and Agile Processes in Software Engineering*, Alghero, Sardinia, Italy.

Baskerville, R., Ramesh, B., Levine, L., Pries-Heje, J., & Slaughter, S. (2003). Is Internet-speed software development different? *IEEE Software, 20*(6), 70-77.

Beck, K. (1999). *Extreme programming explained: Embrace change*. Reading, MA: Addison-Wesley.

Bjerkenes, G., & Bratteteig, T. (1995). User participation and democracy: A discussion of Scandinavian research on system development. *Scandinavian Journal of Information Systems, 7*(1), 73-98.

Boehm, B. (1988). A spiral model of software development and enhancement. *IEEE Computer, 21*(5), 61-72.

Boehm, B. (2002). Get ready for agile methods, with care. *IEEE Computer, 35*(1), 64-69.

Boehm, B., Egyed, A., Kwan, J., Port, D., & Madachy, R. (1998). Using the WinWin spiral model: A case study. *IEEE Computer, 31*(7), 33-44.

Boehm, B., & Turner, R. (2003). *Balancing agility and discipline: A guide for the perplexed.* Boston: Pearson Education, Inc.

Brooks, F. (1975). *The mythical man month: Essays on software engineering.* Reading, MA: Addison-Wesley.

Carmel, E., Whitaker, R. D., & George, J. F. (1993). PD and joint application design: A transatlantic comparison. *Communications of the ACM, 36*(6), 40-48.

Clemont, A., & Besselaar, O. (1993). A retrospective look at PD projects. *Communications of the ACM, 36*(4), 29-39.

Cockburn, A. (2002). *Agile software development.* Boston: Addison-Wesley.

Conrad, B. (2003, October 14). *Taking programming to the extreme edge.* Retrieved from http://archive.infoworld.com/articles/mt/xml/00/07/24/000724mtextreme.xml

Dahl, O.-J., & Nygaard, K. (1966). SIMULA: An ALGOL-based simulation language. *Communications of the ACM, 9*(9), 671-678.

Ehn, P. (1988). *Work-oriented design of computer artifacts.* Fallköping, Sweden: Arbetslivscentrum.

Elssamadisy, A., & Schalliol, G. (2002, May). Recognizing and responding to "bad smells" in extreme programming. In *Proceedings of the 24th International Conference on Software Engineering,* Orlando, FL.

Evans, M. W. (1984). *Productive software test management.* New York: John Wiley & Sons.

Extreme Programming Organization. (2002, November 14). Retrieved from http://www.extremeprogramming.org

Fairley, R. (1985). *Software engineering concepts.* New York: McGraw-Hill.

Greenbaum, J., & Kyng, M. (1991). *Design at work: Cooperative design of computer systems.* Hillsdale, NJ: Lawrence Erlbaum Associates.

Grudin, J. (1991). Interactive systems: Bridging the gaps between developers and users. *IEEE Computer, 24*(4), 59-69.

Henderson-Sellers, B., & Serour, M. (2005). Creating a dual agility method: The value of method engineering. *Journal of Database Management, 16*(4), 1-23.

Hirschheim, R., Heinzl, K. K., & Lyytinen, K. (1995). *Information systems development and data modeling.* New York: Cambridge University Press.

Hirschheim, R., Klein, H. K., & Lyytinen, K. (2003). *Information systems development and data modeling: Conceptual and philosophical foundations.* Cambridge University Press.

Howard, D. (2003). Swimming around the waterfall: Introducing and using agile development in a data centric, traditional software engineering company. In *Proceedings of 5th International Conference on Product Focused Software Process Improvement (PROFES 2004)* (LNCS 2675, pp. 138-145).

Iivari, J. (1990a). Hierarchical spiral model for information system and software development. Part 1: Theoretical background. *Information and Software Technology, 32*(6), 386-399.

Iivari, J. (1990b). Hierarchical spiral model for information system and software development. Part 2: Design process. *Information and Software Technology, 32*(7), 450-458.

Iivari, J., Hirschheim, R., & Klein, H. K. (1998). A paradigmatic analysis contrasting information systems development approaches and methodologies. *Information Systems Research, 9*(2), 164-193.

Joosten, S., & Purao, S. (2002). A rigorous approach for mapping workflows to object-oriented IS models. *Journal of Database Management, 13*(4), 1-19.

Kähkönen, T., & Abrahamsson, P. (2004). Achieving CMMI Level 2 with enhanced extreme programming approach. In *Proceedings of 5th International Conference on Product Focused Software Process Improvement (PROFES 2004)* (LNCS 3009, pp. 378-392).

Känsälä, K. (2004). Good-enough software process in Nokia. In *Proceedings of 5th International Conference on Product Focused Software Process Improvement (PROFES 2004)* (LNCS 3009, pp. 424-430).

Klein, H. K., & Myers, M. D. (1999). A set of principles for conducting and evaluating interpretive field studies in information systems. *MIS Quarterly, 23*(1), 67-93.

Knuth, D. E. (1984). Literate programming. *Computer Journal, 27*(1), 97-111.

Liu, L., Pu, C., & Ruiz, D. D. (2004). A systematic approach to flexible specification, composition, and restructuring of workflow activities. *Journal of Database Management, 15*(1), 1-40.

Luqi, P. B., & Zyda, M. (1990). Graphical tool for computer-aided prototyping. *Information and Software Technology, 36*(3), 199-206.

McConnell, S. (1996). *Rapid development: Taming wild software schedules.* Redmond, WA: Microsoft Press.

McKeen, J. D., Guimaraes, T., & Wetherbe, J. C. (1994). The relationship between user participation and user satisfaction: An investigation of four contingency factors. *MIS Quarterly, 18*(4), 427-451.

Murru, O., Deias, R., & Mugheddu, G. (2003). Assessing XP at a European Internet company. *IEEE Software, 20*(3), 37-43.

Myers, M. D. (2003, April 24). *Qualitative research in information systems.* Retrieved from http://www.qual.aucklandac.nz

Nawrocki, J., Jasinski, M., Walter, B., & Wojciechowski, A. (2002, September). Extreme programming modified: Embrace requirements engineering practices. In *Proceedings of the IEEE Joint International Conference on Requirements Engineering (RE'02)*, Essen, Germany.

Paulk, M. (2001). Extreme programming from a CMM perspective. *IEEE Software, 18*(6), 19-26.

Paulk, M. C., Curtis, B., Chrissis, M. B., & Weber, C. V. (1993). The capability maturity model for software. *IEEE Software, 10*(4), 18-27.

Pressman, R. (2000). *Software engineering: A practitioner's approach* (5th ed.). McGraw-Hill.

Ramesh, B., & Jarke, M. (2001). Toward reference models for requirements traceability. *IEEE Transactions on Software Engineering, 27*(1), 58-93.

Reifer, D. J. (2002). How good are agile methods? *IEEE Software, 19*(4), 16-18.

Salo, O., & Abrahamsson, P. (2004). Empirical evaluation of agile software development: The controlled case study approach. In *Proceedings of 5th International Conference on Product Focused Software Process Improvement (PROFES 2004)* (LNCS 3009, pp. 408-423).

Shabe, G., Peck, S., & Hickey, R. (1977, August). Team dynamics in systems development and management. In *Proceedings of the 15th Annual SIGCPR Conference*, Arlington, VA.

Shoval, P., & Kabeli, J. (2001). FOOM: Functional- and object-oriented analysis & design of information systems: An integrated methodology. *Journal of Database Management, 12*(1), 15-25.

Smart, K. L., & Whiting, M. E. (2001). Designing systems that support learning and use: A customer-centered approach. *Information & Management, 39*(3), 177-190.

Sommerville, I. (2001). *Software engineering* (6th ed.). Addison-Wesley.

Turner, R. (2002). Agile development: Good process or bad attitude? In *Proceedings of 4th International Conference on Product Focused Software Process Improvement (PROFES 2002)* (LNCS 2559, pp. 134-144).

Vanhanen, J., Jartti, J., & Kähkönen, T. (2003). Practical experiences of agility in the telecom industry. In *Proceedings of the XP2003* (LNCS 2675, pp. 279-287).

Visconti, M., & Cook, C. R. (2004). An ideal process model for agile methods. In *Proceedings of 5th International Conference on Product Focused Software Process Improvement (PROFES 2004)* (LNCS 3009, pp. 431-441).

Walsham, G. (1995). Interpretive case studies in IS research: Nature and method. *European Journal of Information Systems, 4*(2), 74-81.

Williams, L., & Kessler, R. (2002). *Pair programming illuminated.* Boston: Addison-Wesley.

Wynekoop, J. L., & Russo, N. L. (1995). System development methodologies: Unanswered questions and the research-practice gap. *Journal of Information Technology, 10*(2), 65-73.

Wynn, E. (2001). Möbius transactions in the dilemma of legitimacy. In E. M. Trauth (Ed.), *Qualitative research in IS: Issues and trends* (pp. 20-44). Hershey, PA: Idea Group Publishing.

Appendix A

Development Environment of Factory System

Users	The factory system is used in factories as well as in sales offices and agencies of the division throughout Europe. Practically all (total of 500) employees are end users of the system. Actually, every user or user group has its own tailored system. One user profile consists of at most 3 to 4 users or shifts. Users receive their work tasks from the system automatically at sign-in. Management has read-only rights into the system.
Tools	The system is developed using an application development tool AdWISE (Western Systems Oy, http://www.western.fi/) and an MDBS IV database (Micro Data Base Systems Inc., http://www.mdbs.com/). AdWISE is a three-tier (client, application server, database server) modular architecture consisting of a fourth-generation application description language W, W compiler, and W interpreter. AdWISE supports prototyping and end-user programming, and makes systems efficient, scalable, and platform independent. LAN (local area network) or WAN (wide area network) is used only for data traffic. MDBS IV is an efficient, reliable, fault-tolerant navigational database system used in mission-critical, real-time applications. With these efficient tools, a standard portable computer or personal computer (PC) is sufficient for developing and running the system and production database. The execution environment is usually small enough to enable the use of, for instance, diskless workstations, mobile phones, and PDAs (personal digital assistants) as clients. In addition, the Cognos PowerPlay software package (Cognos Inc., http://www.cognos.com/) is integrated into the system for OLAP, multidimensional analysis, and reporting.
Team	The development team consists of six persons. The key developers have been in the organization since 1990. The number of developers has gradually increased between 1995 and 2000, with the total number now being four. All developers have worked earlier in other units of the group and in different jobs, so they have a wide experience and total view of the activities in the group and the division. It takes about 6 to 12 months for a new employee to become acquainted with the business. In addition to developers, there are two people on the team who are in charge of the user help desk, training, and testing. The business-development manager and the IT manager, responsible for OLAP, multidimensional analysis, and reporting, also participate actively in the development work. Every developer is familiar with the entire factory system and all the code is mutual. In addition, developers are specialized. One developer is in charge of the sales and statistics applications, one of the production applications, and one of the maintenance and procurement applications. One designer is responsible for the database, the system structure, and the working methods, with one of the three other designers participating actively in these tasks.

continued on following page

Appendix B

Development Environment of Communications Application Portfolio

Users	The four programmers use the applications as tools in developing customer software. The applications are used in about 50 customer implementations, with the total amount of end users being about 300 to 350. About 75% of the customers use the extranet, over 50% use content management, and only a few customers use the other two applications. Per customer, the extranet has 10 to 50 end users, content management has 1 to 10, and the other two applications have one to five end users.
Tools	The application portfolio is built in a LAMP environment around a common application server platform Midgard (Midgard Project, http://midgard-project,org/), which is an open-source framework for information management solutions. LAMP originates from the Linux operating system, Apache (Web server), MySQL (database management system), and PHP (programming and scripting language) components. Some additional code is made using Perl, C++, and Java.
Team	The team consists of four persons. Three software engineers, familiar with each other from their university and still in the middle of their occupational studies, started in 2001. All programmers have profound technological knowledge but little experience of the business. One programmer has left the company and has been replaced with another, who is also an acquaintance from school with around 2 months of apprenticeship. A senior consultant makes up the remainder of the team. The unit manager also participates actively in the development work. Management control over the unit is virtually nonexistent. The unit functions like a miniature open-source software-development community with the main reward system being acknowledgement and approval from peers. All code is mutual, but there is one specialist for every application. Everyone is responsible for customer support and other activities of the team. All programmers are located in the same room sitting by the same table, which makes the communication continuous, informal, and easy. Therefore, the team is very much in-line with the overall philosophy of the case company, which is to be a relatively small, nimble, and efficient one that can quickly adjust to changes.

Chapter II

Understanding Agile Software, Extreme Programming, and Agile Modeling

John Erickson, University of Nebraska – Omaha, USA

Kalle Lyytinen, Case Western Reserve University, USA

Keng Siau, University of Nebraska – Lincoln, USA

Abstract

Failure rates for systems development projects are estimated to approach 50% (Hirsch, 2002). In such an environment, a growing number of developers propose the use of so-called agile methodologies as one means of improving the systems developed while simultaneously decreasing failure rates. Agile proponents insist that adherence to The Agile Manifesto will improve the entire systems development process. This chapter begins by describing some of the agile methodologies, follows that with an overview of current research in the area, and closes with thoughts on possibilities for future applied research into the agile methodologies that could provide evidence supporting or disputing the many claims for success emerging from the field.

Introduction

What are the determinants of success, or conversely, failure, regarding information systems deployments? It seems that IT developers and implementing companies have found as many ways to fail as to succeed. The failure rate of systems development projects is estimated to be more than 50% (Hirsch, 2002). Add to that the fact that many traditional development methodologies are extremely complex and difficult to use, the choice of development and implementation methodology can assume critical proportions. Businesses have come to accept the environment as unarguably turbulent, and the (systems) development environment as a subset appears equally unsettled. In such an arena, it might seem an afterthought that one size does not fit all when it comes to choosing a specific development methodology (Henderson-Sellers & Serour, 2004; Merisalo-Rantanen, Tuunanen, & Rossi, 2004). Enter the agile software development approach as a potential solution. Agile systems development has become the flavor du jour of a group of software developers. Extreme programming and agile modeling are two relatively recent and highly publicized (some would say hyped) specific types of agile development approaches.

While there are many claims for the successful use of extreme programming and/or agile modeling (C3 Team, 1998; Grenning, 2001; Manhart & Schneider, 2004; Poole & Huisman, 2001; Schuh, 2001; Strigel, 2001), and the proponents can often be vocal in the extreme regarding the supposed benefits of both (Ambler, 2001b, 2001c, 2002a, 2002b; Beck, 1999), research evidence supporting the claimed benefits is extremely lacking, although recent work has begun to address at least some of the problems (Fruhling & De Vreede, 2006; Holmström, Fitzgerald, Ågerfalk, & Conchúir, 2006). Currently, the only exceptions seem to be research into two areas.

One, although researchers have begun to study extreme programming, most of the research comprises case studies and field or action research conducted by the principal researcher(s) and related as a case or field report. While this exposition does not intend to detract from the value of a well-conducted case study, additional research into the specific details of the purported benefits of extreme programming would lend some much-needed weight to what appears to be a rather anecdotal body of work. Two, a well-established stream of research into pair programming has generated a set of mixed results that in part provide support for at least one core practice of extreme programming.

The body of research into agile modeling appears to be even sparser than that for extreme programming. Case studies, comparative analyses, and experience reports comprise the majority of the scant research in the area, while very few empirical research efforts have been conducted. Other research efforts encompass the agile software development approach as a whole.

This exposition was written to lay bare the state of research in extreme programming and agile modeling, hereafter known as XP and AM respectively. In addition, research into agile software development will be examined. These goals will be accomplished by first briefly presenting the details of agility, XP, and AM. A literature review for the approaches follows. The chapter then identifies gaps in the literature and proposes possible areas where future study would benefit both research and practice. Finally we conclude the chapter.

Agility, XP, and AM Agility

Agility

Agility is often associated with such related concepts as nimbleness, suppleness, quickness, dexterity, liveliness, or alertness. At its core, agility means to strip away as much of the "heaviness" commonly associated with traditional software development methodologies as possible to promote quick response to changing environments, changes in user requirements, accelerated project deadlines, and the like. The reasoning is that the traditional established methodologies are too set, and often too full of inertia, that they cannot respond quickly enough to a changing environment to be viable in all cases, as they are often marketed to be.

The Agile Manifesto was composed by several XP leaders, promoters, and early adopters and outlines the principles embodied in software and system agility (Lindstrom & Jeffries, 2004). Agile methodologies attempt to capture and use the dynamics of change inherent in software development in the development process itself rather than resisting the ever-present and quickly changing environment (Fowler & Highsmith, 2001). Among the agile methodologies are XP, crystal methodologies, SCRUM, adaptive software development, feature-driven development (FDD), dynamic systems development, and AM.

Information systems development has generally followed a prescribed pattern or process over the past 40 years. Depending upon the specific methodology, the process has assumed many different names, each comprising unique steps. For example, systems developers have proposed the systems development life cycle, the spiral method, the waterfall approach, rapid application development, the unified process (UP), various object-oriented (OO) techniques, and prototyping, to name just a few (Booch, Rumbaugh, & Jacobson, 1999). Many of these time-tested design patterns have evolved into what are now termed "heavyweight" processes.

An influential trend impacting the systems development landscape is the migration to encompass OO analysis and design methodologies. It seems likely that such a move has come about largely as a response not only to the emerging dominance of OO programming languages, but also due to their growing importance to a number of the more recent agile and lightweight development techniques (Fowler & Highsmith, 2001), such as AM and XP. In a very short time, agile software development methodologies have created large waves in the software development industry.

One of the most-used and best-known goal measurement approaches to assessing system complexity and success emerged from work done at Carnegie Mellon in the late 1980s, culminating in the capability maturity model (CMM; Paulk, 2001). The CMM and more recent CMMI takes a goal measurement approach (Pfleeger & McGowan, 1990) and attempts to measure the maturity of the implementing organizations. Five levels of organizational capability and maturity as related to software development constitute CMM. An example of a truly large and encompassing process, the CMM guidelines for becoming a Level 5 organization consume more than 500 pages of requirements. Even stripped to the bare essentials, the CMM comprises 52 primary goals and 18 key process areas (Paulk).

The unified process, while being an OO analysis and design technique, is considered to be a heavy methodology as well. UP consists of four phases, nine disciplines, approximately 80 primary artifacts, 150 activities, and 40 roles (Hirsch, 2002). Even the most optimistic developer looking at UP for the first time would not call it a lightweight or agile methodology. However, Hirsch also provides an experience report consisting of two cases detailing how UP could be modified to be more agile.

The proliferation of development methodologies notwithstanding, it appears that the vast majority of these approaches can be condensed into (or at a minimum contain) four critical steps: analysis, design, coding, and testing

(ADCT). Essentially, AM and XP fit into the ADCT paradigm by breaking the process into very small steps, with each step including the critical analysis, design, coding, and testing elements.

Extreme Programming

XP encompasses four values and four basic activities. The four basic values are communications, simplicity, feedback, and courage. The four basic activities are coding, testing, listening, and debugging (Beck, 1999). According to Beck, these values and activities lead to the 12 core practices of XP: the planning game, small releases, metaphor, simple design, testing, refactoring, pair programming, collective ownership, continuous integration, 40-hour week, on-site customers, and coding standards (Beck; Jeffries, 2001; Wake, 1999).

Beck (1999) presented the primary details and advantages of the approach. According to Beck, XP essentially means to "embrace change." Beck began his exposition by proposing that the basic problem facing software development is risk. Jeffries (2001) proposed that extreme programming is a discipline of software development based on the values of simplicity, communication, feedback, and courage. XP works by bringing the whole team together in the presence of simple practices, with enough feedback to enable the team to see where they are and to tune the practices to their unique situation." Simply put, XP is the coding of what the customer specifies, and the testing of that code is done to ensure that the prior steps in the development process have accomplished what the developers intended. No unforeseen or anticipated tools or features are engineered into the process because XP is oriented toward producing a product in a timely manner. The idea behind XP is that if features are needed later in the development process and the customer notifies the development team at that point in time, the developers need not worry about these features at the present. Needless to say, this represents a vast departure from the normal software development process, in which all requirements (and we naturally suppose these requirements to include features) must be specified up front. This can easily turn into a nightmare since the user requirements can often be seen as dynamic and changing rather than static and set.

According to Turk, France, and Rumpe (2004), XP's values, activities, and practices are quite interrelated, with a relationship structure as follows. Underlying the principles and practices are the basic assumptions that support the XP process. They go on to conduct a more thorough examination of the XP process and connect it to the core beliefs expressed in *The Agile Manifesto*.

Many professionals have proposed different types of modeling or development processes to use with XP, and also to inform developers regarding new concepts in program development (Ambler, 2001a, 2001-2002; Fowler, 2001; Lindstrom & Jeffries, 2004; Palmer, 2000; Willis, 2001). For example, AM, the unified modeling language (UML), UP, and FDD have been used with XP. Since processes used to develop code require modeling, AM as related to XP has been developed. Ideally, modeling techniques help communicate to the entire development team the specifics of a particular design. It appears that the modeling techniques used for AM are as diversified as there are software development scenarios or ideas on the use of XP since UML and UP are essentially modeling techniques.

More recently, Fruhling and De Vreede (2006) have conducted research into extreme programming. They used all of Beck's 12 core practices, some fully and others modified, in developing an emergency response system. Their research aimed at operationalizing the 12 core practices, and certainly indicates that while more focused effort is being expended to investigate the claims of extreme programming success, there is still much to do in terms of measuring that success.

Finally, claims for success abound. For example, Lindstrom and Jeffries (2004) claim that "[t]eams using XP are delivering software often and with very low defect rates." That is great news if it can be verified. Does research show that defect rates are lower in XP-based or other agile methodologies? Rather than simply presenting case studies as examples, documented trends indicating that lower rates are a result of agile practices is necessary before the world will accept the claims as truth. Furthermore, given that some research indicates that pair programming (an XP practice) is not economically viable (Müller & Padberg, 2003), then do lower error rates offset lower productivity?

Agile Modeling

Ambler (2001a) describes agile modeling as "a practices-based software process whose scope is to describe how to model and document in an effective and agile manner." Naturally, the question then arises, does (and if so, how) AM apply to project development executed in an agile development

(XP) environment? Ambler (2001-2002) goes on to develop and explain AM's "core and supplementary principles": simplicity, iterative development, robustness, incremental releases, staying on task, producing a quality product, creating models and the accompanying documentation only as necessary, multiple models, fast and clear feedback on the latest changes to the model(s), and discarding models and documentation that go back more than just a few iterations.

What can be gleaned from the XP approach and applied to AM is the perception that the XP core practices, rather than consisting of isolated ideas about how to create better systems, are quite closely interrelated and interdependent. Essentially, to take the XP approach means to abandon many of the practices that many developers have come to hold dear as critical necessities to systems development. However, since XP merely develops systems, the analysis and design of those systems must also be considered. To do that, developers must model, and to analyze and design effectively for an XP development environment, they should therefore model with an eye toward XP. In other words Ambler (2001a) is in essence proposing that in order to best exploit the benefits of XP, developers should use agile modeling as a lead-in to XP.

XP developers have taken two diametrically opposed perspectives to XP-based systems development (Ambler, 2001-2002). One group proposes that the use of an up-front modeling tool such as UML is necessary to successfully capture and communicate critical system architectures (Armano & Marchesi, 2000). Opposed to the more traditional UML modelers are those who promote the use of UML or other modeling tools only occasionally or simply for graphical representations of the system under development (Willis, 2001). Those developers propose that UML is too complex and heavy to be truly useful in an agile environment. There is some evidence to indicate that UML is indeed complex (Erickson & Siau, 2003, 2004; Siau & Cao, 2001; Siau, Erickson, & Lee, 2002). Furthermore, UML is now even more of a heavy tool with the current move to UML 2.0.

AM basically creates some common ground between the two camps by proposing that developers communicate system architectures by applying AM's core practices to the modeling process (Willis, 2001). This seemingly incompatible marriage of XP practices to UML-like modeling techniques represents the basis of AM. This melding requires two things. First, if a modeling approach is to successfully approximate XP in terms of core practices, then the model must be executable in that it can easily be converted into code and represent to a large extent the functions and features required in the final system. Second, in the context of XP, any models developed must be testable. Recall that developers test more or less continuously in the XP paradigm. This means that, contrary to the common use of UML as merely a tool to draw diagrams, UML in the AM paradigm must be utilized to its fullest extent, and even extended so that the models are executable and testable (Ambler, 2001-2002). Two different tools extending the capabilities of UML into the AM arena have been developed or are currently under development: Argo/UML, from the University of Hamburg, and a petri-net creation named Renew (Ambler). Ambler insists that as these tools move into more mainstream use, the potential advantages of the agile modeling approach combined with extreme programming should become clear.

Extreme Programming and Agile Modeling and Methodology Literature and Research

XP Research

The literature for XP, as previously noted, can generally be split into two separate streams. First, there is a good number of case studies or experience reports that covers the XP approach as a whole, and second, there are research efforts related to one or more of the core practices associated with XP. The experience reports tend to claim success for adopting one or more of the XP practices for specific projects, but offer little in the way of success measures. Since the case studies and experience reports generally involve XP in its entirety, they will be discussed first, and the research related to the core practices second.

XP Cases and Experience Reports

The C3 Team (1998) at Chrysler adopted XP's simplicity value for its compensation system development effort. The team insisted that the project could not have been done in the required time by using the traditionally applied waterfall method. The team found itself behind in implementing a difficult system and discovered that XP lent itself to what they were trying to do. However, the case description does not include much detail in terms of resistance to change that moving to XP might have caused among developers, or other problems encountered that could have been attributed to the XP methodology.

Iona Technologies found that code maintenance and software reengineering were best accomplished by implementing practices they later found to be part and parcel of XP (Poole & Huisman, 2001). They at least partially adopted 11 of the 12 core XP practices, failing only to go to a 40-hour week, and noting that they lacked the courage to try at that point. They also were a bit reluctant about adopting pair programming, noting that many of their programmers were hesitant about trying it.

Schuh (2001) details another in-trouble project that was saved by implementing XP practices. The development team at ThoughtWorks was far behind schedule, working on requirements collected by the previous consultant while the customer had changed the specification and had a rigid delivery date requirement. The project team in this case also partially implemented 11 of 12 XP practices, except for going to a 40-hour week. The team was also hesitant about adopting pair programming.

Another experience report indicates that the team used a traditional "Big Design Up Front" methodology for software development projects (Grenning, 2001). Of the core XP practices, only metaphor was not adopted. Again, while some detail of problems was provided, there is no clear way to discern whether the problems were related to XP or simply part of the process.

XP Core Practice Research

XP education has received its share of attention, meaning that as more industry-based development projects move toward adopting at least some of the XP practices, there has been increasing pressure on university computer science programs to adopt teaching pedagogies with XP embedded. Table 1 indicates some of the investigations and lists the metrics used in the research.

Williams and Kessler (2001) conducted an experiment in pair programming in which they found that traditional postsecondary programming education conditioned students to work alone, and that simply telling them to begin working together does not necessarily result in improved programs, that is, programs with fewer code errors (dependent variable), that are relatively

Table 1. Partial listing of XP core practices research

Author(s)	Type of Study	Independent Variable(s)	Dependent Variable(s)	Results	Threats/Issues
Williams & Kessler	Experiment	Pair vs. nonpair	Code errors	Pair superior	Many; see p. 17
Müller & Padberg (2003)	NPV model simulation	Pair programming and test-driven development	NPV	Lower NPV for XP practices	Simulation vs. real world
Kuppuswami, Vivekanandan, Ramaswamy, & Rodrigues (2003)	Simulation	All XP core practices (effort for each practice individually)	Total effort	Using XP decreases total effort	Simulation vs. real world
Alshayeb & Li (2005)	Field measurement of development project	Changes; growth in class names during project execution	SDI (system design instability)	Refactoring and error fix negatively correlate with SDI	Interaction effects and variables not measured in the study

run-time efficient (dependent variable), and so forth. However, they also noted that once the solo-approach mold was broken, the improvements in finished code were measurable. This is supported by other classroom-based research (Erdogmus & Williams, 2003; Hedin, Bendix, & Magnusson, 2005; Williams & Upchurch, 2001).

In settings outside the classroom, research attempting to assess the benefits of XP and/or its core practices has also been conducted. Aiken (2004) provides support for XP's pair programming practice, noting that although the commonly listed benefits proved attractive to potential adopters, implementing pair programming remained an extremely challenging task. Newkirk and Martin (2000) illustrated via their case study a common problem with software development problems that XP is suited to address. They noted that once the first iteration (of the product) was successfully developed, tested, demonstrated, and delivered to the customer, within a 50-hour window, and according to XP practices, the customer then changed the requirements and added 11 stories to the project. In their view this provided support for the XP approach since, they claimed, a heavy methodology would not have been able to easily incorporate the changes requested.

Müller and Padberg (2003) is one of the few empirical research efforts that dispute the claims made for XP. The authors created an economic model that output a net present value (NPV) of software development projects. The results indicate that, using the XP core practices of pair programming and test-driven development, and comparing with a traditional heavier methodology, the end-product NPV was smaller for the XP-based project than for the traditional project.

Karlsson, Andersson, and Leion (2000) provide an accounting of their experiences regarding the implementation of XP practices at Ericsson, focusing on incremental releases, which they called "daily builds." According to the authors, the project benefited greatly from using the daily-build approach. However, since daily builds also imply daily testing and rigorous attention to coding standards, the implementation effort at Ericsson proved quite challenging.

Agile Modeling and Agile Methodology Research

Agile modeling research appears to be extremely scarce. Armano and Marchesi (2000) adapted UML to an XP-like software development project using what appears to be a combination of the spiral model and UP. The project team committed to weekly builds and refactoring, and used UML to represent user stories via use cases. The team actually developed a tool they called UMLTalk (UML and SmallTalk) to aid in the project development effort.

Other research efforts involve holistic approaches to agile systems. For example, Fujitsu, the Japanese technology company, appears to be one of the early adopters of agile methodologies. It developed an agile tool named Agile Software Engineering Environment (ASEE) as early as 1993 (Aoyama, 1998). The company found itself attempting to complete a software development project from multiple distributed locations and saw the need for an agile approach to solve its problems. The tool was Web-based and enabled releases of software at 6-month intervals for 4 years.

Manhart and Schneider (2004) found that Daimler-Chrysler's embedded software effort for busses and coaches was moving toward a "cautious extension of agile process improvement" after adopting a few (four) agile principles. The development methodology in use was CMM, and the culture appeared to be fairly resistant to change. However, the authors end with a call for more empirical evidence supporting the claims of agile methodologists.

As previously noted, Hirsch (2002) reported on the successful adaptation of RUP for two small projects. Noteworthy of Hirsch's experience report is that the RUP agile adaptation worked best for small projects of 1- to 4-years duration and small development teams of three to eight people. This appears to be a recurrent theme of agile methodologies in general and XP specifically.

Abrahamsson, Warsta, Siponen, and Ronkainen (2003) compared nine different agile methodologies and found that most teams covered different portions of the common development sequence (ADCT) with little or no reasoning as to why they took their specific perspective. Abrahamsson et al. further noted that most agile methodologies did not "offer adequate support for project management." They recommend a focus on quality over quantity, which interestingly enough is a mantra of many of the agile proponents, and end with a comment that empirical research is quite limited.

Potential Areas for Research
Related to XP, AM, and Agile Methodologies

The recurring themes in XP research seem to revolve around XP's pair programming practice. Evidence supporting the idea of pair programming is mounting, and while practitioners conditioned with the heavier approaches such as CMM or RUP tend to resist embracing pair programming, it seems that educators are moving to incorporate pair programming into computer science curricula.

In the case of pair programming research, the effects of programming in pairs are measured against programming individually. Generally, the number of code errors was one measure of differences between programming in pairs vs. programming individually, with some sort of regression testing suite used to assess the errors. Possible confounding variables include such elements as ambient noise (i.e., from other cubicles); using electronic collaboration systems such as MS Net Meeting to collaborate; the physical placement of the computer, keyboard, monitor, and so forth; personal incompatibilities between the two programmers in a given pair; and confusion or ambiguity regarding the role of the person not physically coding. The size of the program to be written is also likely to play a role in the relative success or failure of pair programming as it naturally does with other approaches to coding. Programmer experience is also likely to affect the outcome of pair programming research.

The development and execution of the testing suite, which implies some interaction effects between pair programming and another core XP practice, will also complicate research in this area. At least a few of these moderators or confounders and the threat(s) they pose to validity can be controlled in experimentation, but pair programming in practice should also be examined as part of research in this area.

A relatively large number of experience reports regarding adoption of some the XP practices exist, but hard, empirically based economic evidence is lacking. Many of the case studies and experience reports indicate that most, if not all, XP core practices were successfully adopted. The practice most commonly not adopted was the 40-hour week.

Most experience reports also mentioned that they were already practicing the planning game. XP's planning game can be compared to developing user requirements in a more traditional systems analysis and design development approach. Many of the older and heavier analysis methodologies have well-established evidence regarding the importance of this step. The time spent in analysis could be one analyzable metric, though there are many threats to validity by using time as a variable. In addition, the output of planning can also be measured if artifacts, such as UML diagrams, for example, can be standardized and compared across groups.

Added to the above problem is another that continually plagues developers: that of the aptness of the system. In other words, the planning game might be wonderfully executed along with other cores practices, and the result might be exemplary, but due to changes in the requirements from the outset, the system developed might not be the "correct" system. Of course that problem is endemic to systems development in general, but since XP proponents claim that the approach is superior, then fewer instances of building the wrong system should be evidenced. As to research in this area, while planning is critical and essential for success, it remains to be seen as to whether the specific XP approach is more beneficial than other more standard approaches to developing user requirements.

Companies or organizations using the heavier methodologies typically had trouble adopting incremental releases because of the implications that core practice has for several other core practices: simple design, testing, refactoring, and continuous integration. These core practices appear to be closely related since, for example, a daily build means that the testing suite must also be ready daily, which in turn has implications for continuous integration and refactoring. Research into these core practices will nevertheless be necessary if the overall approach is to be accepted by the mainstream.

If pair dynamic programming is used, the coding-standards core practice means that developers must agree up front on the conventions used for naming classes as well as, for example, on a host of other coding practices. A coding standard in the end means that someone looking at a code segment cannot tell which team member wrote it. This should be something that programmers do for all projects, but sadly it is not. Research should be implemented that compares practice with recommendation in both the traditional and XP areas. However, this instance also highlights once again the difficulties of examining XP's core practices individually: the likelihood that interaction or correlation between and among other core practices will be possible and even probable. In this case, the coding-standards practice is related to and could be affected by pair programming and development of the test suite, just to name two, and there are likely to be other interactions as well.

The efforts of Kuppuswami et al. (2003) represent a pioneering effort in XP research. They used a process model simulation to vary the level (in labor) of XP's core practices one at a time to judge the effect upon total effort for the project. They found that increasing effort (independent variable) into XP core practices reduced the total effort (dependent variable) needed to create the system, although interactions and other moderating effects were not discussed at great length. While the research provides some support for

XP practices, field verification of the simulation is definitely indicated and would be very beneficial.

Other empirical efforts to study XP, in total or just its core practices, are quite limited as well. Williams' numerous and varied studies along with a few others (Alshayeb & Li, 2004; Müller & Padberg, 2003) are the primary exceptions in this area of research. Agile modeling is almost totally unstudied, and any research into the methodology would be an improvement over the current state of affairs. The models themselves could be used as the measures of the efficacy of the methodology, although assessing models as to their relative "goodness" or "badness" is at least somewhat subjective and a possible threat to the validity of research conducted in that manner. The study of agile methodologies appears to be unorganized and, for want of a better word, random.

Conclusion

From a research-based perspective, it appears the research community, practitioners, and educators might benefit from a more structured approach to the study of XP. The bulk of the existing research appears focused on validating the overall XP approach, which is probably, or perhaps arguably, satisfactory if one is only concerned with the macro perspective of XP as a whole. However, since the proponents, as noted previously, seem to universally accept the 12 core practices as integral and necessary parts of XP, then it would seem logical to empirically examine the efficacy of each of the 12 core XP practices separately if we want to examine what it is that makes XP successful (or not). In other words, do we want XP to remain a "black box" and simply accept that it works? Other than pair programming, incremental releases, and at most a few of the other core practices, many of the others remain relatively unstudied, at least in an XP environment.

As to XP specifically or agility in general as approaches to systems development, there is anything but unanimous agreement that there is really anything new. Merisalo-Rantanen et al. (2004) conducted a case study and concluded that XP is really nothing new, but simply a repackaging of old (though arguably useful) techniques for developing systems. Turk et al. (2004) also indicate that the benefits to be gained from adopting agile methods are not realized if the underlying assumptions are not met.

Confounding factors could also cause problems with research in this direction. For example, what are the differences between the prescribed approaches (heavy or light) and practice, and what effect do these differences or gaps have on the success of the various development efforts? Also, if we begin looking at the efficacy of the 12 core XP practices separately, that opens up the possibility of interaction effects. In other words, it appears that a number of the core practices are obviously related to one another—pair programming and collective ownership, for example. If that is the case for obvious and even for nonobvious relationships, then what effect upon the overall success of the project, and ultimately the methodology, does strict adherence to the rules of a prescribed approach for one core practice have if other practices are glossed over or even not used for whatever reason? Perhaps the 12 core practices of XP could even be grouped together into related areas, such as actors (participants in the development effort), technology, structure, and process, and studied from that perspective.

Another potentially critical issue facing software developers and researchers alike is that of software standards. In light of ISO and other standards imposed by governments, implementing organizations, or other regulatory bodies, the quest to render development methodologies more agile by cutting away or eliminating some of the overhead could become difficult or even virtually impossible since the artifacts of development often become a large part of the documentation requirements.

Theunissen, Kourie, and Watson (2003) looked at the potential adaptability of agile software methodologies with regard to ISO/ISE 12207:1995, among other standards. They found that XP in particular could be used to satisfy many of the standard's requirements and developed a set of guidelines for potential users. However, research should be executed regarding whether the guidelines have been successfully adopted and used in practice.

There is no—and likely will never be—an easy fix for these problems. There are no magic solutions (Germain & Robillard, 2005). In addition, from the extremely high failure rates commonly associated with system development efforts (estimates range from 50 to 75%), it appears that there should be ample room for improvement in development efforts, whatever shape or form they take. Agile modeling and extreme programming represent a possible step in the right direction if developers have the courage to commit people and resources to the effort and pain involved in managing the changes that will inevitably occur as a result. However, organizations must also practice caveat emptor and clearly state that they cannot embrace a particular development methodology simply because a person or group of people says that it is good.

The proponents of AM and XP have expressed themselves quite clearly and forcefully on the subject of agile modeling and programming, and, judging from the current bleak and stony landscape of systems development, it appears that they are correct. Those that know (industry insiders and researchers) simply point to statistics that back up their claim that many of the processes we have used and are using to develop information systems are broken and likely unrepairable. When 60% of a typical system's O&M budget goes toward Band-aiding the results of inadequate analysis and development, when two thirds to three quarters (depending upon whose statistics you wish to use) of information systems developed can be considered failures in that they do not provide the functionality required, when we have all been taught to build throwaway systems that can often be obsolete before projects are completed, and when we systematically exclude from development those whom we are building the system for, then it is indeed time to take a step back, look at the mirror, and say, "Just what is wrong with this picture?"

References

Abrahamsson, P., Warsta, J., Siponen, M., & Ronkainen, J. (2003). New directions on agile methods: A comparative analysis. In *Proceedings of the 25th International Conference on Software Engineering* (pp. 244-254).

Aiken, J. (2004). Technical and human perspectives on pair programming. *ACM SISOFT Software Engineering Notes, 29*(5).

Alshayeb, M., & Li, W. (2005). An empirical study of system design instability metric and design evolution in an agile software process. *Journal of Systems and Software, 74*, 269-274.

Ambler, S. (2001a). *Agile modeling and extreme programming (XP)*. Retrieved March 31, 2005, from http://www.agilemodeling.com/essays/agileModelingXP.htm

Ambler, S. (2001b). *Debunking extreme programming myths. Computing Canada, 27*(25).

Ambler, S. (2001c). Values, principles and practices equal success. *Computing Canada, 27*(10).

Ambler, S. (2001-2002). *The principles of agile modeling (AM)*. Retrieved March 31, 2005, from http://www.agilemodeling.com/principles.htm

Ambler, S. (2002a). Agile development best dealt with in small groups. *Computing Canada, 28*(9).

Ambler, S. (2002b). Know the user before implementing a system. *Computing Canada, 28*(3).

Aoyama, M. (2000). Web-based agile software development. *IEEE Software*.

Armano, G., & Marchesi, M. (2000). A rapid development process with UML. *Applied Computing Review, 18*(1).

Beck, K. (1999). *Extreme programming explained: Embrace change*. Boston: Addison-Wesley.

Booch, G., Rumbaugh, J., & Jacobson, I. (1999). *The unified modeling language user guide*. Boston: Addison-Wesley.

Cockburn, A., & Williams, L. (2000). The costs and benefits of pair programming. In *Proceedings of XP 2000 Conference*, Sardinia, Italy.

C3 Team. (1998). Chrysler goes to "extremes." *Distributed Computing*.

Erdogmus, H., & Williams, L. (2003). The economics of software development by pair programmers. *The Engineering Economist, 48*(4), 283-319.

Erickson, J., & Siau, K. (2003, April). UML complexity. In *Proceedings of the Systems Analysis and Design Symposium*, Miami, FL.

Erickson, J., & Siau, K. (2004, December). Theoretical and practical complexity of unified modeling language: A Delphi study and metrical analyses. In *Proceedings of the International Conference on Information Systems*.

Extreme modeling Web site. (2005). Retrieved January 25 from http://www.extrememodeling.org/

Fowler, M. (2001). *The new methodology*. Retrieved March 31, 2005, from http://www.martinfowler.com/articles/newMethodology.html

Fowler, M., & Highsmith, J. (2001). *The agile manifesto*. Retrieved March 31, 2005, from http://www.agilemanifesto.org/

Fruhling, A., & De Vreede, G. (2006). Field experiences with extreme programming: Developing an emergency response system. *Journal of Management Information Systems, 22*(4), 39-68.

Germain, E., & Robillard, P. (2005). Engineering-based processes and agile methodologies for software development: A comparative case study. *Journal of Systems and Software, 75*, 17-27.

Grenning, J. (2001). Launching extreme programming at a process intensive company. *IEEE Software*.

Hedin, G., Bendix, L., & Magnusson, B. (2005). Teaching extreme programming to large groups of students. *Journal of Systems and Software, 75*, 133-146.

Henderson-Sellers, B., & Serour, M. (2004). *Creating a dual agility method: The value of method engineering.* Manuscript submitted from publication.

Herbsleb, J., & Goldenson, D. (1996). A systematic survey of CMM experience and results. In *Proceedings of ICSE-18* (pp. 323-330).

Hirsch, M. (2002). Making RUP agile. *SIG Programming Languages*.

Holmström, H., Fitzgerald, F., Ågerfalk, P., & Conchúir, E. (2006). Agile practices reduce distance in global software development. *Information Systems Management, 23*(3), 7-18.

Jeffries, R. (2001). What is extreme programming? *XP Magazine*. Retrieved March 31, 2005, from http://xprogramming.com/xpmag/whatisxp.htm

Karlsson, E., Andersson, L., & Leion, P. (2000). *Daily build and feature development in large distributed projects.* Limerick, Ireland: ISCE.

Kuppuswami, S., Vivekanandan, K., Ramaswamy, P., & Rodrigues, P. (2003). The effects of individual XP practices on software development effort. *ACM SIG Software Engineering Notes, 28*(6).

Lindstrom, L., & Jeffries, R. (2004). Extreme programming and agile software development methodologies. *Information Systems Management, 24*(3).

Manhart, P., & Schneider, K. (2004). Breaking the ice for agile development of embedded software: An industry experience report. In *Proceedings of the 26th International Conference on Software Engineering*.

Merisalo-Rantanen, H., Tuunanen, T., & Rossi, M. (2004). *Is extreme programming just old wine in new bottles: A comparison of two cases.* Manuscript submitted for publication.

Müller, M., & Padberg, F. (2003). On the economic valuation of XP projects. In *ACM SIGSOFT Software Engineering Notes: Proceedings of the 9th European Software Engineering Conference held jointly with 11th ACM SIGSOFT International Symposium on Foundations of Software Engineering, 28*(5).

Newkirk, J., & Martin, R. (2000). *Extreme programming in practice.* Minneapolis, MN: OOPSLA.

Palmer, S. (2000). *Feature driven development and extreme programming.* Togethersoft Corporation.

Paulk, M. (2001). Extreme programming from a CMM perspective. *IEEE Software, 18*(6), 19-26.

Pfleeger, S., & McGowan, C. (1990). Software metrics in a process maturity framework. *Journal of Systems and Software, 12*(3), 255-261.

Poole, C., & Huisman, J. (2001). Using extreme programming in a maintenance environment. *IEEE Software.*

Schuh, P. (2001). Recovery, redemption, and extreme programming. *IEEE Software.*

Siau, K., & Cao, Q. (2001). Unified modeling language (UML): A complexity analysis. *Journal of Database Management.*

Siau, K., Erickson, J., & Lee, L. (2002, December). Complexity of UML: Theoretical versus practical complexity. In *Proceedings of the Workshop on Information Technology and Systems (WITS)*, Barcelona, Spain.

Strigel, W. (2001). Reports from the field: Using extreme programming and other experiences. *IEEE Software.*

Theunissen, W., Kourie, D., & Watson, B. (2003). Standards and agile software development. In *Proceedings of SAICSIT* (pp. 178-188).

Turk, D., France, R., & Rumpe, B. (2004). *Assumptions underlying agile software development process.* Manuscript submitted for publication.

Wake, W. (1999). *Introduction to extreme programming (XP).* Retrieved March 31, 2005, from http://xp123.com/xplor/xp9912/index.shtml

What is refactoring? (2005). Retrieved March 13 from http://c2.com/cgi/wiki?WhatIsRefactoring

Williams, L., & Kessler, R. (2000a). All I really need to know about pair programming I learned in kindergarten. *Communications of the ACM, 43*(5).

Williams, L., & Kessler, R. (2000b). *The effects of "pair-pressure" and "pair-learning" on software engineering education.* Presented at the Conference of Software Engineering Education and Training.

Williams, L., & Kessler, R. (2001). Experimenting with industry's "pair-programming" model in the computer science classroom. *Computer Science Education.*

Williams, L., Kessler, R., Cunningham, W., & Jeffries, R. (2000). Strengthening the case for pair programming. *IEEE Software.*

Williams, L., & Upchurch, R. (2001, October). Extreme programming for software engineering education. In *Proceedings of the ASEE/IEEE Frontiers in Education Conference*, Reno, NV.

Wills, A. (2001). *UML meets XP.* Retrieved March 31, 2005, from http://www.trireme.com/whitepapers/process/xp-uml/paper.htm

Chapter III

Adaptation of an Agile Information System Development Method

Mehmet N. Aydin, University of Twente, The Netherlands

Frank Harmsen, Capgemini, The Netherlands

Jos van Hillegersberg, University of Twente, The Netherlands

Robert A. Stegwee, University of Twente, The Netherlands

Abstract

Little specific research has been conducted to date on the adaptation of agile information systems development (ISD) methods. This chapter presents the work practice in dealing with the adaptation of such a method in the ISD department of one of the leading financial institutes in Europe. The chapter introduces the idea of method adaptation as an underlying phenomenon concerning how an agile method has been adapted to a project situation or vice versa in the case organization. In this respect, method adaptation is conceptualized as a process or capability in which agents holding intentions through responsive changes in, and dynamic interplays between, contexts and method fragments determine an appropriate method for a specific project situation. Two forms of method adaptation, static adaptation and dynamic adaptation, are introduced and discussed in detail. We provide some insights plus an instrument that the ISD department studied uses to deal with the dynamic method adaptation. To enhance our understanding of the observed practice, we take into account two complementary perspectives: the engineering perspective and the socio-organizational perspective. Practical and theoretical implications of this study are discussed.

Introduction

Despite the best endeavors in the area of information systems research and practice, the effective use of information systems development methods (ISDMs) remains an issue on both academics' and practitioners' agendas (Iivari, Hirschheim, & Klein, 2001). In the 1980s and 1990s, the rationales behind structured, brand-named ISDMs, the so-called conventional methods, were being questioned as being IT oriented, complex, rigid, and inappropriate for postmodern forms of organizations whose distinctive character was to be adaptable to continual change (Sauer & Lau, 1997). Recently, agile—denoting "having a quick resourceful and adaptable character" (Merriam-Webster Online, 2003)—ISDMs, agile methods in short, have appeared as a solution to the long-standing problems related to conventional methods.

This chapter is mainly concerned with the adaptability of agile methods (i.e., the extent to which a method is to be adapted to the project situation at hand or vice versa) yet points out the need for further research in order to understand other distinctive aspects of agile systems' development and to make sense out of the dispersed field of agile methods (Abrahamsson, Warsta, Siponen, & Ronkainen, 2003). As we shall see later on, many studies concerning the effective use of ISDMs adopt the notion of adaptation but use different terms or concepts in their theoretical constructs, for example, "method fragment adaptation" in Baskerville and Stage (2001), "scenario use" in Offenbeek and Koopman (1996), "method tailoring" in Fitzgerald, Russo, and O'Kane (2000), "situational" or "situated method engineering" in Harmsen, Brinkkemper, and Oei (1994) and Slooten and Brinkkemper (1993), "context-specific method engineering" in Rolland and Prakash (1996), and "method engineering" in Siau (1999).

Two limitations with these studies have motivated us to carry out this research. First, the existing studies use different perspectives and provide countervailing arguments for the notion of adaptation. Second, the proposed models appear to be limited to theoretical arguments and need empirical findings to support their arguments. More precisely, as Fitzgerald, Russo, and O'Kane (2003, p. 66) state, "little research has been conducted to date on method tailoring specifically." This observation is particularly true for agile methods.

Our research addresses these two limitations and illustrates the working practices in a large-scale IT department dealing with the adaptation of an agile method, dynamic systems development method (DSDM), as elaborated later on, in different project situations. Besides the description of the observed practice, this chapter argues the need for a multitheoretic lens combining the engineering and the socio-organizational perspectives, and uses it to elaborate the notion of adaptation in agile systems development. Similar to the research approach adopted by Fitzgerald et al. (2003), this chapter inductively draws lessons from agile method adaptation in practice rather than tests hypotheses defined in advance. In doing so, the chapter provides valuable insights for both practitioners and academics concerning the effective use of agile methods in large-scale IT departments.

The structure of the chapter is as follows. First, the motivation behind the research has been outlined in this section. The remainder of the chapter consists of three key sections: (a) a review of related research, (b) the conduct of this research, and (c) discussions and conclusions of the research.

Background

Given that the existing explanations concerning method adaptation are fragmented and countervailing, we need a framework in which to organize the previous research relevant to method adaptation. Such a framework will also help us indicate the focus of this chapter. Before introducing the framework, we will clarify our interpretation of key terms such as method and method adaptation, and their usages in this chapter.

The first term is method. As various definitions of method exist, we use its simplest yet broadest meaning; as such, it refers to "an explicit way of structuring one's thinking and actions" (Jayaratna, 1994). While new methods are promoted as a panacea for well-publicized ISD failures, old ones have been criticized for being rigid, comprehensive, and built upon the idea that a method can be used for all projects, which brings on a "one-size-fits-all" issue. In fact, a fundamental problem still remains that methods, irrespective of their preferred features (agility, state-of-the-art knowledge foundations), by nature involve certain thinking and often prescribe certain actions for ISD. The subject matter at hand addresses this one-size-fits-all issue and aims to deal with how an ISD method is adapted and can be supported so that the resulting method, the so-called situated method, fits a project situation. The idea behind a situated method is that any prospective method to be used for a development project is subject to certain adjustments because of the fact that the method is limited to its preferred thinking and prescribed actions for ISD that cannot fully accommodate the uniqueness of a project situation. In this regard, such adjustments are needed for the method along with a premise that the resulting method can provide a well-suited means for ISD and in turn reduce the risk of its failures.

The fundamental assumption about method existence in IS development states that as long as IS development takes place, a method must be present in the development of an IS, and human actions and thinking involved in IS development are purposely structured to achieve certain goals. This assumption follows from the definition of method (an explicit way of structuring one's thinking and actions). Accordingly, method has two essential functions in ISD: (a) the function that purports certain effects on human thinking (such as augmenting, facilitating, and structuring), for which we use the term intellectual, punctuating the strategic orientation of a method fragment, and (b) the function that purports certain effects on human behaviour (i.e., supporting, automating), for which we use the term procedural, emphasizing the operational orientation of a fragment.

These two functions are intimately intertwined because it is granted in the field of the philosophy of mind that certain kinds of human behaviour (e.g., purposive human behaviour) cannot be truly isolated from associative cognitive models (schemata) in the human mind. A number of researchers in the domain of method engineering, including Siau (1999), have studied a psychological perspective on method adaptation. Similar argument can be made for techniques and tools that can treat methods, tools, and techniques as methodical means and/or intellects—that is, possessing certain viewpoints on thinking about IS and ISD, which supposedly aim to support practitioners for effective and efficient development of an IS. As methodical means the focus of methods is on their practical use; as such they can support practitioners' actions during ISD. Besides practical use, one can look at implications in the context of human thinking related to development activities. In this sense, methods interact with their users and such interactions can be seen as hermeneutic—that is, interpretative—processes (Introna & Whitley, 1997). This has to do with mutual understanding, and augmenting and structuring their way of thinking about IS and ISD. Methods are just instruments, but along with this interaction they have another role and posses certain intellects and even aim to convey certain thinking about IS and ISD. As such, methods have their own reasoning mechanism or understanding of whatever methods are supposed to do. It is this understanding that gives methods power to influence the way of thinking held by the practitioners in the course of action during ISD.[1] It is this fact that gives methods a special role in the development of human intellects.[2] Having presented the meaning of method, we can turn our attention to adaptation.

The second term is adaptation. The term adaptation simply implies "a modification according to changing circumstances" (Merriam-Webster Online, 2003). Since its significance might vary, for the purpose of this chapter, we further define method adaptation as a process or capability in which agents through responsive changes in, and dynamic interplays between, contexts, intentions, and method fragments determine an appropriate method for a specific project situation. With this definition we aim to stay at an abstract level that will allow us to organize related previous research. Before explaining the terms in the definition above, two key perspectives concerning method adaptation are introduced.

As noted in Baskerville and Stage (2001), existing studies related to method adaptation follow one of two key perspectives: the engineering perspective representing the positivist views of natural science, and the socio-organizational perspective representing interpretative views of social science (see Table 1). The former is of interest to the school of method engineering, emphasizes the structural aspects of the method, and usually employs contingency-based models for method adaptation. The latter appears to be concerned with better understanding of how a method and its components are invented on the

Table 1. Framework for organizing previous research relevant to method adaptation

	The Engineering Perspective	The Socio-Organizational Perspective
Agent	Method engineers as dominant actors	An interplay between people, including project managers, method engineers, developers, and end users, involved in a project
Contexts	Factor-based characterization of context	Emerging context in ISD setting
Method Fragment	Coherent and structured parts of a method	Innovated, unstructured fragments separated from a prescribed method
Process/Intention	Static and dynamic use of factors mediated by an intention, often in terms of risk and success factors	An ill-structured, complex organizational phenomenon

fly and are actually used in an emerging work setting, and this is reflected in the body of knowledge contained in the socio-organizational literature (Baskerville & Stage).

These two perspectives adopt different levels of abstraction for method adaptation (Iivari, 1989). The engineering perspective stays at a conceptual level where the main focus is on models of the real or empirical world rather than the real world itself (Harmsen, 1997). In comparison, the socio-organizational perspective looks into the empirical world and tries to understand method adaptation in practice, examining real, concrete development processes. The empirical study of Fitzgerald et al. (2003) presented how method adaptation had been carried out in the Motorola organization at various levels. The authors distinguished three adaptation levels: the industry, the organization, and the project. Our focus in this chapter is on method adaptation at the project level.

Prescribed vs. Emerging Context

The term context refers to a collection of relevant conditions and surrounding influences that make a project situation unique and comprehensible (Hasher & Zacks, 1984). The complexity of context as a subject has been acknowledged by many scholars, including Akman and Bazzanella (2003). Andler (2003) argues that relevant discussions on this subject in philosophy evolve from its narrowest meaning about the consideration of texts in linguistics to its broadest meaning, something to do with "situated cognition,"[3] which is invariably situated, as elaborated in the field of pragmatism. In particular, a traditional view of the notion of context suggests that contexts are preexisting and stable environments that perhaps include unobservable factors that cause agencies to behave in partly unpredictable ways (Rogoff & Lave, 1984). This view appears to be akin to what Andler calls the optimistic claims stating that for all classes of cognitive tasks and processes, there is a uniform context matrix, whatever the features or factors are granted, such that for all situations in the class, the outcome of any process in the class is determined by the values taken by the matrix in the situation.

This is often contrasted with the contemporary view that asserts that all contextual regularities, conditions, and any other relevant features are assumed to be dynamically activated and accomplished in the situation (Linell & Thunqvust, 2003). Context has also been studied as a central notion in human decision making. Pomerol and Brézillon (2001) illuminate the dynamics of context and the employment of reasoning for practical decision making. Practical decision making, as discussed by Pomerol and Brézillon (2004), is reminiscent of naturalistic decision making, an adopted orientation in this work.

Different kinds of context are introduced with a duality character (Schegloff, 1992) such as immediate or proximate contexts. These include features pertaining to actual surroundings in situ vs. distal or mediate contexts that cover background knowledge, cognitive frames, or assumptions about ongoing, upcoming, or even a priori activities relevant in situ. Another distinction is made between so-called primary and secondary context, the extent to which influencing characteristics are stable (Pomerol & Brézillon, 2001). In relation to this duality character, Andler (2003) defends a "mixed model of inquiry," which combines rationalist reliance either on fact or principles with a consideration for appropriateness to the situation at hand. This is indeed where the pragmatics view of context stands and of which several accounts are proposed. Mey (2003), for instance, advocates this view and argues that ambiguity is inherent in contextualization, decontextualization, and recontextualization through which one may effectively marginalize certain agencies and their legitimate interpretations by virtue of an institutionally embedded context.

But what does context then include? Or to say it differently, what is included and excluded in this contextualizing? Brézillon and Pomerol (2002) suggest focusing on the dynamic of context rather than things included and excluded in contextualizing and propose three types of knowledge for contextualizing: external, a part of knowledge not used in a specific situation at the moment contextualizing occurs; contextual, a part of knowledge relevant for contextualizing; and proceduralized, a part of knowledge invoked, structured, and effectively situated in contextualizing. Perhaps a more provocative question would be who excludes what, and on whose premises? These questions have to do with the roles of agencies in this contextualization. Andler (2003) states that:

... the ultimate goal of a general theory of context would be an account for regularities, if any, which can be observed in the effects of context on cognitive process. If there are indeed such regularities, the context problem, relative to the class of situations and processes at hand, has an in-principle solution, consisting in refining and otherwise modifying the state space. (p. 354)

Human agency is central to contextualization. In connection with this work, of course, method fragments are also considered during this contextualization. However, exclusion of the agency and method fragments is in effect when the context is framed and reframed along with the cognitive structure and processes (Piaget, 1983). After successive approximation, this eventually leads to an appropriate context under consideration in which the decision is made. Accordingly, cognitive structures change through the process of adaptation by assimilation and accommodation. This is boldly marked in the radical constructivism along with the principle stating that the function of cognition is adaptive and serves the agency's framing or organizing of the experiential world, not the discovery of an objective ontological reality (Glasersfelds, 1997). We employ the ideas of contextualizing, framing, and appropriation in relation to the very notion of context.

Interested readers can see the elaborations of existing models or views characterizing the context in which an IS development takes place (Lyytinen, 1987). Both the perspectives discussed above use various kinds of factors to understand the context. Even though the proposed list of factors in the domain of method engineering is supposed to be lengthy, it is apparent that social and organizational issues are not the focus of attention. The socio-organizational perspective, however, does put more emphasis on social and organizational elements of the context. In addition, this perspective considers context as an emerging ISD setting rather than as a prescribed project situation.

Structured vs. Innovated Method Fragments

Both perspectives use the concept of fragments. From the engineering perspective, a method fragment is a description of an ISDM or any coherent part thereof. It is usually prescribed and structured in terms of fragment properties (Harmsen, 1997). Conversely, the socio-organizational perspective gives more attention to those fragments that are distinct from a prescribed method. This perspective sees fragments as follows: "Under [this] concept, each systems development project is a moving pastiche of miscellaneous parts; bits of external methodologies, internal methods, innovative, unique techniques invented on-the-fly, etc." (Baskerville & Stage, 2001, p. 18). To differentiate between the two meanings of this concept, we consider there to be two types of fragments. We use the terms structured and unstructured fragments to refer to the meanings in the engineering and socio-organizational perspectives respectively.

Fragments can be principles, fundamental concepts, products to be delivered, activities needing to be performed, job aids—techniques, tools, hints, tips—to be used, and so forth. Some of them are essential to the ISD approach. The term ISD approach, and we adopt the definition of Iivari et al. (2001), refers to a high-level description of the method including the goals and the guiding principles, and the beliefs, fundamental concepts, and principles of an ISD process.

Fragments can be related to aspects of the method, such as the way of thinking, modeling, working, controlling, and supporting (Wijers, 1991). We are interested in those fragments related to the way of thinking, and to some extent to the way of modeling and working. The terms principles and assumptions used in published methods often refer to this kind of fragments (Turk, France, & Rumpe, 2005). These are called strategic fragments in that they have strategic orientations or effects on the way of thinking on ISD and IS and reflect intellectual function of the method. They are concerned with, for instance, modeling aspects and scope, development strategy, and deployment strategy. As such, they are often referred to as building blocks of scenarios or a planned approach in literature (e.g., Slooten & Hodes, 1996).

So, we identify the following fragments and corresponding decision variables related to several aspects such as the modeling aspect, design-development aspect, and user-engagement aspect. Consequently, we have the following.

Strategic fragments that are related to modeling aspects:

- **Modeling scope (the boundary of the target system and dimensions):** The extent to which the approach considers the tracing of several perspectives such as functional, information, process, organizational, and operation (e.g, Curtis, Kellner, & Over, 1992)
- **Approach orientation (the orientation of the problem-solving system):** Problem or solution orientation and social aspect (technical-administrative or social-organizational; see Offenbeek & Koopman, 1996)
- **The analysis starting point (knowledge acquisition strategy):** Current situation or future situation (direct acceptance of user requirements; the actual system as a starting point, possibly from the point of view of the old system; determining information requirements from scratch, starting from perspective of the object system)
- **Reuse (design) strategy:** Using a reference (architecture) model, a new architecture, or a combination of both

Strategic fragments that are related to design-development aspects:

- **Dividing strategy:** Increment strategy (how to partition the problem and/or solution space)
- **Realization strategy:** The way to realize a number of increments—at once (no subsystem), concurrently (parallel), overlapping, or consecutively (subsystems are developed one after another, incrementally)
- **Development strategy:** Linear, overlapping, throwaway, keep-it prototyping, evolutionary, or reverse engineering
- **Delivery strategy:** The way to deploy a solution in an organizational setting—big bang (at once), incremental, evolutionary

Strategic fragments that are related to stakeholder-engagement aspects:

- **Validation strategy:** Immediate acceptance, definition of norms, and test cases by means of which assessment takes place or whether the chosen solutions meet the requirements; prototyping; validation by usage

- **Engagement strategy:** Based on the interaction model of Offenbeek and Koopman (1996) and in particular on the user engagement (degree of user involvement and responsibility)

Agents Leading Method Adaptation with an Intention

An agent is an actor with one role or more in a method adaptation process. The socio-organizational perspective does not specify any specific roles in that process, yet the emphasis is on the practical interplay between people at work. The socio-organizational perspective considers the method adaptation process as "an ill-structured, complex socio-organizational phenomenon" (Baskerville & Stage, 2001, p. 14). Anthropology is referred to as a potential reference discipline to study such a process, and Agar's (1986) practical ethnography and its four major units of analysis are used to explain how the process develops in practice.

The engineering perspective regards method engineers as the dominant actors in method adaptation. Their role is to carry out the process leading to a tailored method, that is, a method that is adapted to the project context at hand. Such a process usually employs contingency-based models. Offenbeek and Koopman (1996) discuss the limitations of 17 contingency-based models that have been proposed for determining an appropriate approach for an IS development project. As they note, the factors taken into account in these models can be numerous, or limited to certain IS views and used in a static manner. That is, these models ignore possible bilateral interactions between the context, characterized by the factors, and the approach, and further lack dynamic interactions among the factors. Offenbeek and Koopman propose the concept of a dynamic fit between context and approach as a solution to the static use of contextual factors, the approach, and the corresponding method fragments. They state, "To a certain extent the dominant actors cannot only choose their approach but also their context, whether by definition or by intervention, that is by deliberately changing the context" (p. 257). It is important to note that both the context and the approach are subjects for adaptation, and a form of mediating construct is needed to facilitate this adaptation process. Such a construct is here called an intention and has been referred to using different terms in the various models proposed for method adaptation; see, for instance, risk in conventional contingency-based models as listed in Van Offenbeek and Koopman (1996), success in Harmsen et al.

(1994), goal in Baskerville and Stage (2001), and mediating factors in Slooten and Brinkkemper (1993). We consider the intention as an indication of what drives the agents while carrying out method adaptation.

In the dictionary (Merriam-Webster Online, 2005) and everyday language, the term intention is synonymous with volition, purpose, and significance, and indicates "a determination to act in a certain way." Other derivations and uses of the term appear as intent, intentionality, doing with an intention, or doing something intentionally. To ground explanations concerning their differences would require a long philosophical treatise that belongs to the philosophy of mind, but the treatment of intention and intentionality in Bratman (1987) and Morison (1970) is relevant to our subject. The treatment of the terms intention and intentionality should be separated as the former has been articulated in relation to action, planning, and practical rationality (Bratman), and the latter is proposed in phenomenology, a particular school of thought in the philosophy.

Intention is considered a state of mind (what it is to intend to do something) and a characteristic of action (having an intention to do something or doing something intentionally). Intentionality derives from the Latin verb intendere, which means to point to or to aim at, and Brentano (1838-1917) accordingly characterized the intentionality of mental states and experiences as their feature of each being directed toward something. Intentionality in this technical sense then subsumes the everyday notion of doing something intentionally: An action is intentional when done with a certain intention, that is, a mental state of aiming toward a certain state of affairs. One of the most comprehensive expositions of the term intention is in the work of Michael Bratman. His treatment reveals complexity and the essence of its characteristics and functions along with two forms (future and present directed[4]). Bratman (1987) extensively discusses his account in relation to planning theory and agent rationality, for which we cannot condense the body of literature he employs in a few pages. The forms and kinds of intention he proposed, however, are especially useful for characterizing the agency action in method adaptation. Upon deeper examination of the idea of intending to act, which channels a future-directed form of intention, or having an intention to act, which is present-directed action, he contends that intentions are neither desires nor beliefs but plans, and that plans have an independent place in practical thinking. One of the central facts about intentions essential for this work is that they are conduct-controlling pro-attitudes and serve as inputs for further practical reasoning.

The Conduct of the Study

Research Objective

During our case-study investigation in an organization, we explored, described, and analyzed the work practices dealing with method adaptation without limiting ourselves to a specific perspective. To frame our research scope, we formulated our goal as to investigate how an agile method is adapted to different project contexts in a large-scale IT department. By using the constructs elaborated in the previous section, this goal statement could be formulated as follows: to investigate the ways through which a method engineer and a project manager together adapt dynamically both structured and unstructured fragments of an agile method to different contexts at the project level. We especially looked into the early stages of the systems development process where the adaptation process appeared to be more essential and more transparent in the organization investigated.

Research Method

The research approach adopted in this study is that of an interpretive field study. Many researchers, including Fitzgerald et al. (2000) and Sauer and Lau (1997), have also used this research approach for the study of method use in practice. It has been suggested as an appropriate research method for explorative and descriptive types of research and, according to Klein and Myers (1999, p. 69), "interpretive research does not predefine dependent and independent variables, but focuses on the complexity of human sense making as the situation emerges; it attempts to understand phenomena through the meanings that people assign to them."

The field research was conducted in the form of a research project in the organization and carried out by a research team consisting of people from both the university and the case organization. The Appendix summarizes the characteristics of the research method applied, such as the use of multiple study stages, various sources of knowledge, an iterative process of data analysis (Walsham, 1995), a collaborative style of the research team's involvement, "engaged" data gathering (Jones & Nandhakumar, 1993), and the use of different feedback mechanisms for the validity of the data analysis. One can see that the mentioned characteristics are indeed related to the principles of interpretive field research (Klein & Myers, 1999). (Due to a space limitation, we could not further articulate the relations between the characteristics and the principles, but as an example, notice that the use of various sources of knowledge is related to the principles of multiple interpretations, suspicion, and contextualization.)

Introducing the Case Organization

The organization we investigated is one of the leading financial institutions in Europe and operates in a dynamic business environment. One of the global strategic business units, Consumer and Commercial Clients (C&CC), focuses exclusively on services to individual clients and small- to medium-sized businesses. The Netherlands Business Unit (BU) is one of the five BUs under C&CC. IT Development is one of the departments within the Netherlands BU and employs 2,000 people involved in systems development projects. Such a large IT department was chosen because it enabled us to investigate method adaptation in various project contexts.

It is worth noting that the organization has considerable experience of ISD-method use. The organization's identity goes back 10 years to the merger of two organizations, both of which were used to using conventional methods. One of them had been using a method developed in house, and the other a brand-named method. Until the introduction of an agile method, just 2 years ago, there had been a lot of effort put into achieving a standard method influenced heavily by previous development procedures, processes, and templates.

About the Agile Method: DSDM in a Nutshell

Dynamic systems development can be considered an agile method because it has the ability to be adaptable to a variety of development situations (Abrahamsson et al., 2003). In the United Kingdom and in Benelux countries, DSDM, which is supported by a consortium of over 600 organizations, has become the de facto market standard. The method strongly emphasizes the concepts of suitability and adaptability; DSDM will be, to a certain extent, suitable for a project or an organization, and is adaptable if not completely suitable.

For the purpose of this research, we have considered three components of DSDM: its underlying philosophy (captured in nine principles), its framework (stages, activities, products), and its essential techniques (Aydin & Harmsen, 2002). In practice, each of these components can be applied separately, and subsets of the components can also be applied on their own. The principles of the method are active user involvement, frequent delivery of products, iterative and incremental development, an empowered team, fitness for business purposes, reversible changes, requirements at a high level, testing throughout the life cycle, and a cooperative approach. The DSDM framework suggests a complete project approach that includes key phases, products, and roles that should be customized according to the project situation (see Table 2 for the examples of product and process fragments of the DSDM used at the business-study level). Modeling techniques are not included in DSDM since they are often a part of modeling tool sets that are not themselves part of the method. In this way, DSDM is highly adaptable: It is possible to use fully fledged DSDM, but individual techniques or just the terminology are still valuable on their own. To this end, an instrument called a suitability filter is available in the manual (DSDM Consortium, 2003). The filter considers the critical success factors for DSDM and the characteristics of projects that will make DSDM especially effective. Each potential project should be judged individually using the filter. If the project provides a good match with the filter, then DSDM can be considered as a suitable method. If the criteria results are not satisfied, then the method can be modified.

Table 2. Examples of product and process fragments of the DSDM used at the business-study level

Business-Study Level		
Product Fragments		**Process Fragments***
Main Products	*Models*	
Business area Definition Outlined prototyping plan System architecture definition	Business functions Data/ relationships/rules Business events Business scenarios Business architecture System locations	Visionary Ambassador user(s) Project manager

*Note: *Only roles-related fragments are provided here. See the complete list of fragments in DSDM (2000).*

Important DSDM techniques are time boxing, facilitated workshops, prioritization, and prototyping. Time boxing refers to setting a deadline by which a predefined objective must be met rather than describing when a task must be completed. To prioritize requirements of the system, the MoSCoW technique is used; the term is an abbreviation for the phrase "must have, should have, could have, and want to have, but won't have this round." We assume that the concepts of facilitated workshops and prototyping are known. For more details of DSDM, one should refer to the DSDM Consortium document (DSDM Consortium, 2003).

The Situation at Hand

Recently, DSDM has become the method of choice for all information system development projects in the department. The main motivation for this decision was to ensure time-to-market systems development in order to achieve substantial product and process improvements, and to use one terminology in all projects. The DSDM implementation in the department focused on coaching project managers in adapting the method in the organization and at project levels with the help of experts. The experts, known as coaches, had extensive project experience and were subject-matter experts in DSDM use. They coached project managers on how to make better decisions on the suitability of DSDM and on the degree of adaptation DSDM would require for each project. Basically, there were two essential, important roles in DSDM adaptation: the project coaching role and the project management role. The DSDM coaches assisted project managers in adapting DSDM to their project context, whereas project managers were fully responsible for the project execution. They were the final decision makers in terms of the use of DSDM fragments.

Case-Study Procedure

The field research consisted of three stages: the preliminary study stage, the actual research stage, and the posterior study stage (see Table 3).

We conducted the research in cooperation with a sponsor and a method engineer from the case organization. The sources of knowledge were, in this empirical setting, informants, direct observations, and documents. Since

Table 3. Summary of the case-study procedure

Time	Stage	Event / Activity	Objectives	Involved People
Jan 2002	Preparation	Field-study preparation	• Uncovering all aspects of the phenomena that has been studied so far in the literature regarding two theoretical views • A high-level description of the research method is to be used	Academics, including one primary investigator and three senior researchers (professor, assistant professor, and a subject-matter expert)
Feb	Preliminary Study	Conducting, codifying, analyzing, and reporting interviews	Explained in the Appendix	Explained in the research-method section
May	Preliminary Study	Discussion of the reflections of interview results within the organization		All research team members and method engineers
May	Preliminary Study	Determining research scope, and research design variables	Explained in the Appendix	Research team members
June	Actual Study	• Second-round interviews • Third-round interviews • Direct observation • Artefacts analysis (route maps, instruments such as the ESRL, advice documents, etc.) See Appendix for other activities	Explained in the Appendix	Explained in the Appendix
June July Sept	Actual Study	Three checkpoint meetings	• Validation of findings • To agree on the level of abstraction and degree of generalization • To agree on the depth and breath of the research scope	Explained in the Appendix
July Sept	Actual Study	A number of discussion meetings with a broad audience	Explained in the Appendix	Explained in the Appendix
Nov	Actual Study	Closing up and writing a draft version of the case protocol	To document findings in a scientific way	Academics
Dec 2002-Mar 2003	Posterior Study	Several iterations for the case protocol	Quality improvement by peer reviews	Academics (internal and external)
Dec 2002-June 2003	Posterior Study	Follow-up communications with the organization	Explained in the Appendix	Explained in the Appendix
Sept 2003	Posterior Study	Informal meetings	Monitor the evolving practice specific to method adaptation	Explained in the Appendix

the information needed was partially available in the organization, the team concluded that several rounds of formal and informal interviews, direct observations in the form of attending meetings, and in-depth documentary analysis were the most appropriate ways to collect data. Essentially, three rounds of interviews were conducted, each at a different level of detail in different forms, with different informants (i.e., embedding different levels and roles). In some interviews, a list of questions was used to ensure that all the important subjects were covered, but at the same time, room was left for emerging issues (see the Appendix for the interview questions and other details of the research method used).

In this interpretive case research approach, we preferred engaged data-gathering methods to distant ones as they allowed us to gain rich insights into method adaptation (Jones & Nandhakumar, 1993). However, some limitations of this approach have been identified. One of the problems, as frequently cited in the IS literature (e.g., Klein & Myers, 1999), was the difficulty in controlling the interactions between the researchers and the participants, especially in a large IT development department. Another problem was the level of abstraction needed and the degree of generalization achieved. To assess these problems, the research team members organized three checkpoint meetings in which up-to-date research findings were discussed and the scope of the future stages of the research determined. In these meetings, the depth and breadth of the research scope was elaborated and found to be satisfactory for all the parties involved in this research. Another type of feedback mechanism, used to check the validity of the analysis, was to present and discuss the research findings with other interested parties in the case organization. This involved 12 method engineers, six project managers, one change manager, one chief domain architect, and two quality-assurance leaders. The feedback from such a broad audience was useful to justify our findings.

Major Findings

We identified static and dynamic method adaptations as two distinct ways of carrying out method adaptation in the department. Next, we describe each of them separately.

Static Method Adaptation

Static method adaptation refers to selecting and assembling structured fragments based on a predefined set of criteria. In the case organization, we found that the type of development or target environment (i.e., the technical infrastructure, or the platform an application will be designed and built upon) and the nature of the solution (i.e., a packaged or a custom-made application for business change; Gibson, 2003) were two of the dominant factors used in static adaptation. Static method adaptation resulted in several route maps. A route map is an established plan prescribing which structured fragments should be used in a project. Examples of route maps are packaged solutions and component-based development (CBD; Dahanayake, Sol, & Stojanovic, 2003). These route maps have some similarities with the form of process landscapes as described in Backlund, Hallenborg, and Hallgrimsson (2003). In the event of choosing a route map for a project, the project manager could see only the relevant structured fragments, including stages, activities, products, techniques, and modeling tools for that project. It was interesting to note that the relevance of principles and essential DSDM techniques were not specified as part of these route maps. This point encouraged us to investigate how unspecified fragments have been adapted in practice, and so we needed to look at the second adaptation level.

Dynamic Method Adaptation

The second way for method adaptation, which we refer to as dynamic method adaptation, takes place during the process of developing an agile system. In this way, the role of the coaches is essential in order to adapt both structured and unstructured fragments to the contexts or vice versa. In practice, the coaches in the department were facilitating project managers to choose, modify, or innovate fragments for each project. As a consequence, we decided to focus on coaching activities and studied the means used in method adaptation. Figure 1 summarizes the key activities performed by the coaches. Two decisions had to be made in this coaching activity diagram: whether to use DSDM or not (in the suitability analysis), and whether to adapt or directly use parts of DSDM (in the adaptation analysis). Note that the output of characterizing the project was used with both decision points. Next, we discuss the ways and means that can be used to characterize a project.

Figure 1. Overall coaching activities regarding method adaptation

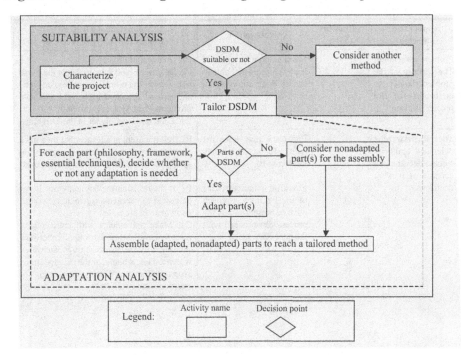

We noted that coaches were using an instrument, the so-called Extended Suitability/Risk List (ESRL), for characterizing a project. During the early stages of DSDM use in the department, the coaches had used the questions in the original DSDM suitability filter (DSDM Consortium, 2003). Later, as they gained experience with them, some questions were extended and clarified, and furthermore, for each question, working instructions, measures, useful hints, and tips were added (Table 4).

The ESRL became an instrument that provided a baseline for the written advice to be produced for each project. In our interviews with both the coaches and the project managers, participants emphasized the significance of using the ESRL in method adaptation. They commented on the high relevance of the factors in the ESRL for better understanding the project situation at hand. In the ESRL, the applicability factors are closely related the preconditions and principles that need to be taken into account for the effective use of the method. These, in fact, reflect most of the success or risk factors that are often

Table 4. The extraction from the ESRL

Applicability Factor	Suitability (Y/N)	Explanation	Management Measure (P=Preventive, C=Corrective)
Problem ownership: The identity of the problem holder, or customer for the project, is clear.		Is a champion (proponent/ leader) present and able to ensure that resources are released?	P1. Do not start project. P2. Determine who actually holds the purse strings and who ultimately makes decisions and carries the responsibility. Who will have problems if the system is not built? C1. Look one level higher in the hierarchy.
The end users with the delegated authority to make decisions are capable of making decisions.		End users may have the required authority, but may fail to use it. Essential characteristics of the iterative approach must be present so that the process can proceed with the necessary speed.	P1. Tell the users in advance that they have the authority to make decisions within the specified boundaries and that they must indeed make these decisions. P2. If the decision-making authority is not delegated to users, management must also participate in the team. C1. Make agreements with management regarding availability; for example, questions submitted by the teams must be answered within x days, x hours, or the manager must keep a half an hour free every morning for questions (e.g., 8:30-9:00).

cited in IS literature (Schmidt, Lyytinen, Keil, & Cule, 2001). To clarify the meaning of each factor, the instrument includes further explanations with some follow-up questions and examples (see the Explanation column in Table 4). The instrument basically accepts the following assumption: that the inapplicability of the factors to the context at hand can cause a discord between the preconditions for effective use of the method and the project context. To mitigate the discord and related issues, suggestions are provided in the form of preventive and corrective measures in the instrument (see the Management Measure column in Table 4). These measures indicate the preconditions for the effective use of the method and relate them to the fragments of the method. We noted that the coaches considered the measures as suggestions rather than as directives for method adaptation. They had discussed the appropriateness and applicability of the measures with project managers. The coaches and project managers had discussed the implications of method adaptation in terms of conformance to time and budget (i.e., the degree to which the desired functionality could be realized within an agreed time or budget), and customer satisfaction (the degree to which the project outcomes would fulfill the expectations of the sponsor and users).

Table 5. The extraction from the sample advice

About the Project Context	About the Appropriate DSDM Development Strategy
"If we know that the requirements are almost clear, stable, and that it is hardly possible to prioritize them, that there is no clear user interface, that there is high computational complexity, that the timeline is not clear, and that the resource availability (in terms of developers, end user) is not known, yet the total resources can be fixed, then we would like to know which development strategy is most appropriate and what kind of consequences we may anticipate in the later DSDM phases."	
	About Some Issues Related to Two Techniques of DSDM and Related Risks
	"… as the case indicates, the MoSCoW (a DSDM technique) appears not to be very suitable for this situation due to the difficulty of prioritizing requirements. The same holds for timeboxing, for which there must be a fixed date for the project, or for an increment, or for an iteration. For both anticipated issues there may be some opportunities to use these two techniques in different ways. Indeed, DSDM coaches have had some experience with such ways and they successfully use the philosophies behind MoSCoW and Timeboxing in real projects situations."

Once a coach had used the ESRL and discussed the implications of method adaptation with project managers, they would write down their advice on how best to use the method for a successful system development in the perceived project context. To give a flavor of such advice, we have provided Table 5, and with this we will illuminate the notion of structured and unstructured fragments.

Let us first focus on the advice about the appropriate DSDM development strategy. The recommendation given is closely related to the principle of iterative and incremental development, which simply states that "many increments with iterations is an ideal development strategy for agile systems development" (DSDM Consortium, 2003). Using increments means that a solution can be split into components that are based on prioritized requirements (Slooten & Hodes, 1996). More formally, an increment is a part of the system that is delivered to, and used by, a user before the total system is operational. However, having iterations means that some stages and corresponding activities need to be repeated through incorporating continuous feedback from the user. Such an iterative aspect of a development strategy contributes to the achievement of fitness for business purposes, which is another principle of the method.

The hybrid development process recommended in the sample advice shows how the principle of iterative and incremental development can be adapted to the project context described in Table 4. It suggests that a project manager should realize some increments in an iterative manner, and achieve the rest without iterations (i.e., by applying a linear or waterfall systems development strategy). The term hybrid underscores the mixture of typical DSDM development strategy (iterative and incremental systems development) and a linear development strategy in such a project context.

The other part of the advice regarding issues about two techniques of DSDM and related risks on the one hand addresses structural parts of the method—that is, the techniques MoSCoW and timeboxing—and on the other hand points out an unstructured innovative fragment by noting that "[i]ndeed, DSDM coaches have already experienced such ways and they have successfully used the ideas behind MoSCoW and Timeboxing in such a project context." The innovative fragment here is to use timeboxing in a different way to that prescribed in a given project context. One coach explained how to use timeboxing in a different way:

It is true that you usually use timeboxing when the deadline of a project is known and then you can split a fixed timeline into "boxes," but you can also do it by using budget as a criterion. Namely, if the human resources to be used in your project are known, you can calculate total available human resources in terms of man-hours and then you can convert this into a fixed budget and apply the idea of timeboxing as "budgetboxing."

In fact, we identified many such structured fragments that needed to be adapted and these resulted in innovative fragments in the case organization. However, given the space limitation in this chapter, we have simply presented a few examples of such fragments in this section, and we will discuss their implications in the next section.

Discussion and Conclusion

The findings presented in the previous section show that the two perspectives are complementary and may even be necessary rather than conflicting if one considers adapting both structured and unstructured method fragments for two distinct approaches to method adaptation in a large-scale IT department

Table 6. Characteristics of the static and dynamic adaptations for an agile method in the case organization

Two Ways for Method Adaptation / The Constructs Relevant to This Research	The Static Adaptation	The Dynamic Adaptation
Key Perspectives Applied	The engineering perspective	Both the engineering and socio-organizational perspectives
Levels of Abstraction	The conceptual level	The empirical level
Agent	Only coaches or other method engineers	The coaches and project managers
Contexts	Factor-based characterization of context, characterized by the nature of a solution and the type of development or target environment	Emerging context in an ISD setting, characterized by a set of factors in an instrument
Method Fragment	Only the structured fragments (stages, activities, modeling tools)	Both structured and innovated (unstructured) fragments
Process/Intention	Only adapting the method to the context; the static use of factors with an intention to adhere to the method	Adapting the method to the context or vice versa, with an intention to adhere to time and budget, and achieve customer satisfaction

(see Table 6). In the following, we shall explain this complementary aspect of the two perspectives.

Static Adaptation

As summarized in Table 6, the engineering perspective, embedding the dynamic-fit concept of the contingency paradigm, provides a sound basis to illuminate static adaptation. Indeed, method engineers have been primarily responsible for characterizing a project context and determining which fragments are needed for a project. The chosen fragments, which result in various route maps, are good examples of the models created at the conceptual level. It is rather easy to see that a high degree of method adherence was driving the process for static adaptation. It is also clear in this process that the direction of adaptation is from method to context; that is, method is adapted to context.

Static adaptation helps project managers start with an appropriate route map for a particular project, but it has some limitations on the way to character-ize the context in which the project runs. Namely, as we pointed out before, such adaptation employs a prescribed view of the context by using foreseen and salient contextual factors. This implies that static adaptation at best leads to a kind of a prescribed method by incorporating a priori project-specific characteristics. As we have seen from the present case, a project manager has needed dynamic adaptation to be able to adapt method fragments and context to each other in the course of a project.

Dynamic Adaptation

Similar to static adaptation, dynamic adaptation helps a project manager to adapt the chosen fragments to the context in the project execution. In this adaptation, depending on what the context requires and what the intention is, project managers need to further modify the structured fragments or even innovate new fragments. We shall now consider two types of fragments to illuminate modification and innovation of fragments.

For the former, consider our finding about how the timeboxing technique (setting a deadline by which a predefined objective must be met), which is one of the essential techniques of the method, has been used in some projects. This technique is essential in that it can be used as a means to achieve some of the principles of the method, such as frequent delivery of the system or its parts, or the quick incorporation of feedback from the project stakeholders to the system to be delivered. We have showed that even though the technique (a structured, chosen fragment), at first glance, was not suitable for the project context, the agents strove to accommodate this technique in a special proj-ect context (no timeline was set for a project) and found an alternative way (budgetboxing) to apply the essence of this technique. It was clear that the intention behind this adaptation was partly due to the desire to adhere to the method, and partly to adhere to the philosophy behind the technique.

For the latter, consider our finding about how the principle of iterative and incremental (a structured fragment) development was changed to a hybrid approach (an innovated fragment). We have showed that the hybrid approach was recommended as an appropriate development strategy to the project context as described in Table 5. This means that, on this occasion, the context forced agents (project managers and coaches) to find out an alternative way of using the principle of iterations and increments.

In contrast with static adaptation, dynamic adaptation allows a project manager to adapt the project context to method fragments in the course of a project (adaptation at the empirical level). To explicate this point, we can refer to the Management Measure component of the ESRL tool. This contains some suggestions concerning the ways to change the context. For instance, the inapplicability of a factor related to the user, as presented in Table 4, may require some management measures. These measures in fact indicate how the context might be changed to mitigate the issues possibly faced in order to realize the fragments of the method, which are mainly related to the philosophy component of the method. In this event, the reaction of the agents can be to change the context and/or the fragment. We have seen that the intention that drove the behaviour of the agents was closely related to the desire to conform to time and budget, or to customer satisfaction.

Even though agents do their utmost to mitigate risks and related issues, a project is not risk free, and the agents might be faced with some emerging breakdowns resulting from a discord between the method and the context. These breakdowns may eventually result in risks for the project. Such breakdowns need to be resolved, possibly by innovating new fragments or substantially changing the existing fragments. The socio-organizational perspective helps to illuminate such fragments, pinpoint the root causes of breakdowns, and describe methical and amethodical aspects of the breakdowns (Truex, Baskerville, & Travis, 2000). In addition, this perspective facilitates an understanding of the emerging context in which the resolutions have to be achieved and the fragments invented. In this sense, the ESRL, on the one hand, employs the engineering perspective and helps agents to characterize and adapt the context and fragments. On the other hand, the ESRL accommodates the socio-organizational perspective and helps project managers to make sense of what the emerging context is about and what fragments are being innovated in such a context.

Proactive Role for the Agent Involved in Method Adaptation

An important implication of method adaptation is related to the degree at which an agent is dominant for method adaptation. In fact, the idea of method adaptation asserts that method, context, and the agent are not passive elements in the interplays among them, but purposively intervene in the agent's knowledge about how to handle the construction of the situated method. This implies that we should advance in our thinking about the effect of method in these interplays rather than reducing its meaning to certain aspects and attributes. To show how to advance in thinking, we suggest looking beyond the "frozen" rationale captured and often implicit in the presence of the method, and possibly capture its creator's way of structuring the intended user's (the designer role) thinking and actions. This advanced understanding of method is related to its intellectual function; the practical function is more geared to structuring actions. Most methods are proposed to make use of the practical function of the method, but this is limited in its use and has possibly severe consequences if the agent is unaware of the intellectual function. The consequence can be so dramatic that the agent can become a slave of the method if she or he is not confident about the fragment. Nontechnically speaking, if the agent is not familiar with the method and is forced to use it, then either the agent's thinking or actions are fully captured in the method or severe clashes and breakdowns occur between the agent and method. These often occur at later stages and may cause project failures. This means that the agent should be more proactive in revealing and preventing these breakdowns. Guidance in this research explicates how the agent (like a project manager in the case organization) can be supported in this respect. The role of mediator (like a coach in the case organization) is essential to support the designer in the awareness of limitations of not only the method, but also his or her own fragment. In this regard, we suggest that method should be enacted with its intellectual function so that it will not tell you what and how things should be done but act like an advisor and facilitate the agent in constructing a truly situated method. Implication of this change in method functioning is substantial for its creator. Instead of providing the full-fledged content of a method, the experience of those who use the method should be a starting point for establishing the basis of a method. This idea resembles the method life cycle consisting of several loops (ad hoc approach → best practice → de facto method → de jure method → ad hoc approach) as mentioned in Harmsen (1997).

Method Adaptation in Globally Distributed System Development

Traditionally, systems development activities are colocated and almost no methods are designed specifically for this purpose. All parties are close, so many activities are carried out face to face. However, the trend in practice is changing toward systems that have been developed in a more globally distributed manner. Methods fall short in addressing the challenges of how to conduct globally distributed systems development (GDSD). It is interesting to see how method adaptation deals with differences among parties involved in such settings in terms of ways of thinking (along with culture, laws, language, etc.) or acting (distribution of work, communication and coordination mechanisms, etc.). Not only is distributed global systems development needed in practice, but distributed global method adaptation would also be required. In case the method fails to accommodate globally distributed systems development, we can expect method adaptation would be driven by the context at hand. This suggests that since the method does not address the aforementioned challenges driven by GDSD, people would be forced by the context to come up with a new practice that leads to innovative method fragments. Studying method adaptation in GDSD would provide new insights in understanding the effect of contextual differences on MAP.

Practical Implications

Practical implications of this study are manifold. First, we can argue that two approaches to adaptation—static and dynamic—could be applicable and useful in a large-scale IT department. We especially focus on the dynamic adaptation rather than the static adaptation and emphasize that for the dynamic adaptation, the role of coaches is found to be essential in supporting project managers to make appropriate decisions on the use of method fragments in a specific project context with an intention. This chapter details how such support was achieved in the case organization. Second, it is our contention that an instrument similar to the ESRL, but incorporating up-and-working experiences derived from real projects, might be useful in supporting the agents (the method engineers and project managers) in dynamic method adaptation. This study shows the feasibility, applicability, and usefulness of such an instrument in the context of agile systems development in one of the leading financial institutes in Europe.

One of the implications of this study for academics is that the constructs drawn from relevant research and summarized in Table 1 can provide a solid theoretical ground for future research regarding method adaptation. Notice that in this study we have articulated these constructs and used them to explore the adaptation of an agile method to different project situations in a large-scale IT department (Table 5). For future research, there is an opportunity given by the fact that by using these constructs, one can investigate other agile methods in different organizational settings to further discern the role of the key constructs described in the framework. Another research opportunity related to the proposed constructs is to study the relations between these constructs. Such a study might propose and possibly test a number of hypothetical relations between the constructs for static adaptation and/or dynamic adaptation. Notice that in this study we just give some indications of how these constructs might be related for two types of method adaptation.

Comparison with Other Studies

Regarding the comparison of our findings with relevant studies, we shall comment on the following subjects. First we will discuss the use of a multitheoretic lens on method adaptation. It seems that for studying method adaptation, such an approach is novel in academic circles although the complementary aspect of two perspectives has already been mentioned as a future research topic by Baskerville and Stage (2001). Second, most of the findings about method adaptation, including the Motorola case presented by Fitzgerald et al. (2003), and the cases of Ericsson ERA/RNC and Volvo IT presented by Backlund et al. (2003), are similar to those presented here, but their analysis either stays at the organizational level or focuses on only the static adaptation of other methods. Our work covers both static and dynamic adaptation of an agile method (DSDM). This study considers DSDM as an example of the agile method and shows empirical evidence on the situational appropriateness of DSDM at the project level, which is found to be a missing point in literature (Abrahamsson et al., 2003). A final comment can be made about the distinction between DSDM and other agile methods on method adaptation. Even though other agile methods claim to support method adaptation at the project level, most of them lack clear guidance on how to do this. DSDM includes an instrument aiming at guiding project managers in realizing method adaptation. We have emphasized that such an instrument provided the case organization a good starting point to work on the relevance of the content of the instrument to its own project situation. That is why instead of going into detail about the content of the instrument the organization had used, we have especially focused on its dimensions and the way it had been used in method adaptation.

However, this research also has some limitations. Even though DSDM is an excellent example of an agile method, one has to take into account the limitations of the findings since they are specific to one method and one case organization. Consequently, we have discussed the findings from two perspectives in order to draw lessons inductively rather than generalize them and test previously defined hypotheses.

Conclusion

Based on our experience, we hope that this chapter will encourage other academics to employ two perspectives when investigating agile methods. To realize static and dynamic adaptations as two distinct ways of carrying out method adaptation, organizations can benefit from using a coaching service and instrument as described in this study. We especially emphasize on how dynamic adaptation incorporates two perspectives and has been realized by the help of the coaching service and the instrument used in the case organization. However, while we try to draw the attention of academics to the use of the two perspectives in method adaptation, we cannot ignore the fact that the engineering perspective has had a privileged position in the history of conventional methods. As a consequence, we need to especially increase our knowledge on the use of the socio-organizational perspective in gaining a better understanding of agile methods adaptation.

The research community in which our work is positioned has dedicated research domains (so-called information systems development and method engineering domains) on the subject matter and has a solid body of knowledge. In that sense, our contribution might be regarded as a modest extension of the body of knowledge in these research domains, consisting of further articulation, explication, and establishment of the idea of method adaptation, which refers to the phenomenon about dynamic interplays between a context, an agent, and a method fragment in an information systems development situation. Naturally and essentially, the foundation of method adaptation needs to be established by using existing bodies of knowledge and more empirical studies. It is natural that such a modest extension is needed because the very notion of agent deserves more attention as the heart of method adaptation. It is essentially needed because without this notion, method adaptation lacks its essential feature referring to how the agent in some way adapts her or his knowledge to the context or the other way around. One can argue about where her or his adaptive capability comes from. We all have this capability, which goes beyond the basic discussion of survivability. Whether it is granted or learned, it is this capability that makes the agent aware about what is going on and helps the agent involved in method adaptation in particular to manage intriguing interplays among herself or himself, the context, and the fragment.

References

Abrahamsson, P., Warsta, J., Siponen, M. T., & Ronkainen, J. (2003). New directions on agile methods: A comparative analysis. In *Proceedings of the 25th International Conference on Software Engineering* (pp. 244-254).

Agar, M. (1986). *Speaking of ethnography.* Newbury Park, CA: Sage.

Akman, V., & Bazzanella, C. (2003). The complexity of context: Guest editors' introduction. *Journal of Pragmatics, 35,* 321-329.

Andler, D. (2003). Context: The case for a principled epistemic particularism. *Journal of Pragmatics, 35*(3), 349-371.

Aydin, M. N., & Harmsen, F. (2002). Making a method work for a project situation in the context of CMM. In M. Oivo & S. Komi-Sirvö (Eds.), *Product focused software process improvement* (LNCS 2559, pp. 158-171). Berlin, Germany: Springer.

Backlund, P., Hallenborg, C., & Hallgrimsson, G. (2003, June). *Transfer of development process knowledge through method adaptation and implementation.* Paper presented at the 11th ECIS, Naples, Italy.

Baskerville, R., & Stage, J. (2001). Accommodating emergent work practices: Ethnographic choice of method fragments. In B. Fitzgerald, N. Russo, & J. I. DeGross (Eds.), *In realigning research and practice: The social and organizational perspectives* (pp. 11-27). Boston: Kluwer Academic Publishers.

Bratman, M. (1987). *Intention, plans and practical reason.* Harvard University Press.

Curtis, B., Kellner, M. I., & Over, J. (1992). Process modeling. *Communications of the ACM, 35*(9), 75-90.

Dahanayake, A., Sol, H., & Stojanovic, Z. (2003). Methodology evaluation framework for component-based system development. *Journal of Database Management, 14*(1), 1-26.

Dynamic Systems Development Method (DSDM) Consortium. (2003). *Dynamic systems development method.* Retrieved from http:/www. dsdm.org/

Fitzgerald, B., Russo, N., & O'Kane, T. (2000). An empirical study of system development method tailoring in practice. In *Proceedings of the 8th International Conference on Information Systems* (pp. 187-194).

Fitzgerald, B., Russo, N., & O'Kane, T. (2003). Software development method tailoring at Motorola. *Communications of the ACM, 46*(4), 65-70.

Gibson, C. F. (2003). IT-enabled business change: An approach to understanding and managing risk. *MIS Quarterly Executive, 2*(2), 104-115.

Glasersfeld, E. von. (1997). Piaget's legacy: Cognition as adaptive activity. In A. Riegler, M. Peschl, & A. von Stein (Eds.), *Understanding representation in the cognitive sciences: Does representation need reality* (pp. 283-287)? New York: Kluwer Academic/Plenum Publishers.

Harmsen, F. (1997). *Situational method engineering.* Utrecht, the Netherlands: Moret Ernst & Young Management Consultants.

Harmsen, F., Brinkkemper, S., & Oei, H. (1994). Situational method engineering for information systems projects. In T. W. Olle & A. A. V. Stuart (Eds.), *Methods and associated tools for the information systems life cycle* (pp. 169-194). Amsterdam: North-Holland.

Hasher, L., & Zacks, R. T. (1984). Automatic processing of fundamental information: The ease of frequency of occurrence. *American Psychologist, 39*(11), 1372-1388.

Hutchins, E. (2000). *Cognition in the wild.* Cambridge, MA: The MIT Press.

Iivari, J. (1989). Levels of abstraction as a conceptual framework for an information system. In E. D. Falkenberg & P. Lindgreen (Eds.), *Information systems concepts: An in-depth analysis* (pp. 323-352). Amsterdam: North-Holland.

Iivari, J., Hirschheim, R., & Klein, H. K. (2001). A dynamic framework for classifying information systems development methodologies and approaches. *Journal of Management Information Systems, 17*(3), 179-218.

Introna, L. D., & Whitley, E. A. (1997). Against method: Exploring the limits of method. *Information Technology & People, 10*(1), 31-45.

Jayaratna, N. (1994). *Understanding and evaluating methodologies*. Berkshire: McGraw-Hill.

Jones, M., & Nandhakumar, J. (1993). Structured development? A structurational analysis of the development of an executive information system. In D. E. Avison, J. E. Kendall, & J. I. DeGross (Eds.), *Human organisational and social dimensions on information system development* (pp. 475-496). Amsterdam: North-Holland.

Klein, H., & Myers, M. (1999). A set of principles for conducting and evaluating interpretive field studies in information systems. *MIS Quarterly, 23*(1), 67-93.

Linell, P., & Thunqvist, D. P. (2003). Moving in and out of framings: Activity contexts in talks with young unemployed people within a training project. *Journal of Pragmatics, 35*(3), 409-434.

Lyytinen, K. (1987). Different perspectives on information systems: Problems and solutions. *ACM Computing Surveys, 19*(1), 5-46.

Merriam-Webster online. (2003). Retrieved November 3, 2003, from http://www.m-w.com

Morrison, J. C. (1970). Husserl and Brentano on intentionality. *Philosophy and Phenomenological Research, 31*, 27-46.

Offenbeek, M. A. G. van, & Koopman, P. L. (1996). Scenarios for system development: Matching context and strategy. *Behaviour & Information Technology, 15*(4), 250-265.

Piaget, J. (1983). Piaget's theory. In P. Mussen (Ed.), *Handbook of child psychology*. Wiley.

Pomerol, J.-C., & Brézillon, P. (2001). *About some relationships between knowledge and context: Modeling and using context (CONTEXT-01)* (LNCS, pp. 461-464). Springer Verlag.

Rogoff, B., & Lave, J. (1984). *Everyday cognition: Its development in social context*. Cambridge, MA: Harvard University Press.

Rolland, C., & Prakash, N. (1996). A proposal for context-specific method engineering. In S. Brinkkemper, K. Lyytinen, & R. J. Welke (Eds.), *Method engineering: Principles of method construction and tool support* (pp. 191-208). Atlanta, GA: Chapman & Hall.

Sauer, C., & Lau, C. (1997). Trying to adopt system development methodologies: A case-based exploration of business users' interests. *Information Systems Journal, 7*, 255-275.

Schegloff, E. (1992). In another context. In A. Duranti & A. Goodwin (Eds.), *Rethinking context* (pp. 191-1227).

Schmidt, R., Lyytinen, K., Keil, M., & Cule, P. (2001). Identifying software project risks: An international Delphi study. *Journal of Management Information Systems, 17*(4), 5-36.

Searle, J. (1983). *Intentionality: An essay in the philosophy of mind.* New York: Cambridge University Press.

Siau, K. (1999). Information modeling and method engineering: A psychological perspective. *Journal of Database Management, 10*(4), 44-50.

Slooten, K. van, & Brinkkemper, S. (1993). A method engineering approach to information systems development. In N. Prakash, C. Rolland, & B. Pernici (Eds.), *Information system development process.* Amsterdam: Elsevier Science Publishers B.V. (North-Holland).

Slooten, K. van, & Hodes, B. (1996). Characterizing IS development projects. In S. Brinkkemper, K. Lyytinen, & R. J. Welke (Eds.), *Method engineering: Principles of method construction and tool support* (pp. 29-44). Atlanta, GA: Chapman & Hall.

Truex, D., Baskerville, R., & Travis, J. (2000). A methodical systems development: The deferred meaning of systems development method. *Accounting, Management & Information Technology, 10*, 53-79.

Turk, D., France, R., & Rumpe, B. (2005). Assumptions underlying agile software-development processes. *Journal of Database Management, 16*(4), 62-87.

Walsham, G. (1995). Interpretive case studies in IS research: Nature and method. *European Journal of Information Systems, 4*(2), 74-81.

Wijers, G. M. (1991). *Modelling support in information systems development.* Delft, the Netherlands: Delft University of Technology.

Appendix:
About the Research Method Applied

Research Stages	The Preliminary Study Stage	The Actual Study Stage	The Posterior Study Stage
The Sources of Knowledge and the Techniques Used to Interact with Participants	*Informants:* Six method engineers First round of interviews in the form of semiopen formal interviews	*Documentary analysis:* The organization-wide development method; the existing route maps and related fragments; an instrument (the ESRL) used for method adaptation; templates and actual project documents, including advice documents, project proposals, and systems development plans *Direct observations:* Attending daily meetings of method engineers First round of interviews in the form of open-ended and semiopen (formal and informal) interviews — *Informants:* 12 method engineers Second round of interviews in the form of open-ended and semi-open (formal and informal) interviews — *Informants:* 12 method engineers, six project managers, two portfolio managers, one change manager, two quality-assurance leaders, one chief domain architect	*Informants:* The head of the coaching group and some method engineers
Main Research Focus	• Determining relevant context(s) for the ways in which an agile method is adapted • Gathering perceptions and opinions of method engineers on method adaptation in general	• Identifying and studying the prescribed forms (route maps) of the method • Identifying tailoring drivers behind the prescribed forms • Studying the formulation of structured and unstructured fragments • Exploring, describing, and analyzing working practices and a means that the department uses to deal with the static and dynamic adaptations • Studying the practice for dynamic adaptation in detail	Being up to date on the subject matter
Sample Questions	What do you think about the adaptability of the method (DSDM) to a project situation? What about previous and current practices on method tailoring? How do you go about tailoring it for a specific project? How do you support project managers on this matter? What kind of information do you exchange with project managers?	What do you think about the coaching support (provided or received) for a project? What do you look for and take into account when tailoring the method for a specific project situation? Could you explain the activities and the knowledge used while coaching a project manager? How do you determine the suitability of the method to a project? What do you use for it? What do you do if the prescribed parts of the method do not fit the project context? Do you use any means to characterize a project? What do you think about the instrument (the ESRL)? What about the contextual factors and measures in the instrument? How do you use them? How do you write down your advice on how best to use the method for the project? How do you use the advice in your project? What about the relevance of the instrument and its parts (contextual factors, measures) to the task concerning method adaptation? Are the factors and measures meaningful, comprehensible, and useful for method adaptation?	What has been changed in method adaptation practice so far? Any change regarding coaching support, other working practices, the means, or so forth?

Chapter IV

Matching Models of Different Abstraction Levels:
A Refinement Equivalence Approach

Pnina Soffer, Haifa University, Israel

Iris Reinhartz-Berger, Haifia University, Israel

Armon Sturm, Ben-Gurion University of Negev, Israel

Abstract

This chapter deals with the reuse of models, which assists in constructing new models on the basis of existing knowledge. Some of the activities that support model reuse, such as model construction, retrieval, and validation, may involve matching models on the basis of semantic and structural similarity. However, matching for the purposes of retrieval and validation relates to models of different abstraction levels, hence structural similarity is dif-

ficult to assess. This chapter proposes the concept of refinement equivalence, which means that a detailed model is a refinement of an abstract model. It emphasizes the use of refinement equivalence for the purpose of validating a detailed application model against an abstract domain model in the context of a domain analysis approach called application-based domain modeling (ADOM). We discuss the structural characteristics of refinement operations in object-process methodology (OPM) models, and present an algorithm that detects refinement equivalence.

Introduction

The benefits of applying reuse at various stages of system design and implementation have been widely recognized. The reuse of software components has been addressed for over 40 years, and the idea has been extended to other and more abstract design artifacts, such as design models and specifications (Eckstein, Ahlbrecht, & Neumann, 2001; Kim, 2001; Reinhartz-Berger, Dori, & Katz, 2002; Zhang & Lyytinen, 2001), requirements models (Lai, Lee, & Yang, 1999; Massonet & Lamsweerde, 1997; Sutcliffe & Maiden, 1998), conceptual models (Pernici, Mecella, & Batini, 2000), enterprise models (Chen-Burger, Robertson, & Stader, 2000), method engineering models (Ralyte & Rolland, 2001), and others. When the reusable artifact is a model, the purpose of reuse is to assist in constructing a new model, either within the same domain, or within another domain by analogical reasoning.

Reuse is a major underlying motivation for the emergence of the domain engineering discipline. Domain engineering supports the notion of a domain, defined as a set of applications that use common concepts for describing requirements, problems, and capabilities. The purpose of domain engineering is to identify, model, construct, catalog, and disseminate a set of software or business artifacts that can be applied to existing and future systems in a particular domain. A subfield of domain engineering is domain analysis, which captures and specifies the basic elements of the domain and the relationships among these elements, representing this understanding in a useful way. Domain analysis is, therefore, a discipline that deals with creating reusable models of a domain and reusing these models for creating specific applications.

Reuse environments of models in general, and domain analysis environments in particular, should provide support to at least part of the following activities: (a) construction of reusable models and their storage, possibly in a repository, (b) retrieval of models (or parts of them) that meet the requirements of a developed application, (c) adaptation of the reusable models to the current application needs, and (d) validation of the adapted models. These activities may employ in some cases a model matching operation, which is the focus of this chapter.

In the context of domain analysis, two types of reusable models can be used. One is a generic domain model at a high level of abstraction that has to be specialized in adaptation to the current needs. The second type is a complete and detailed model, whose level of abstraction is the same as that of the application. It may be reused as it is, or modified to the specific needs, but without a change in its abstraction level.

The abstraction level of the reusable model affects the nature of the above discussed activities. First, reusable models of a high abstraction level are constructed by abstracting a collection of domain applications and analyzing their commonalities and variation points. Model matching may be employed for detecting the common aspects of the collection of application models that are being generalized.

Second, the role of a repository is of much importance for low-level reusable models since a large number of these may be stored, and each may include slightly different details. In contrast, high-level domain models specify common aspects of domain applications; hence, a large number of such models is not required.

Third, in general, the retrieval of a model can be either index based or model based. Index-based retrieval uses indices that characterize the models, while model-based retrieval matches an input model (query) given by the user with the models stored in the repository (Mili, Mili, & Mili, 1995). While index-based retrieval is relatively simple and quick, model-based retrieval is more accurate, relying on a higher volume of information rather than on a classification represented by indices. Retrieval of a high-level model is relatively simple due to the low number of models and the clear distinction between them, hence, index-based retrieval is appropriate. Retrieval of a low-level detailed model is more complicated since there may be a number of different models for a given domain, and retrieval seeks the one that matches partial information available about the particular current needs. Model-based retrieval, relying on all the information captured in a model, enables the selection of the model that best fits the user's query. It may use a preliminary partial model or some facts about the modeled domain as an input query, and retrieve a detailed model (or detailed models) that matches the input model.

Fourth, the adaptation of a high-level model to the current needs is an instantiation operation, yielding an application model that should match the domain model. This matching should be verified by a validation activity. The adaptation of a detailed model can be done by modification (which can be controlled through defined variation points) or by integration with other models. Validation in this case should follow the variation points and check that their specified constraints are not violated.

In summary, model matching can be used for the activities of constructing a reusable model, retrieving it, and validating an application model against the reusable one. When model matching is used for retrieval, the expected output is a similarity measure, while when it is used for construction or validation, the focus is on identifying specific matches and mismatches between the models.

This chapter deals with the assessment of structural similarity between two models of a different abstraction level. Soffer (2005) addressed this issue emphasizing its relevance for the retrieval of a detailed model. Here we address the scenario of validating an application model against a domain model. Addressing this scenario, we decided to rely on an existing domain analysis approach in order to relate to concrete details rather than taking a generic view, which might overlook the complexity of the task. The domain analysis approach we use is application-based domain modeling (ADOM; Reinhartz-Berger & Sturm, 2004; Sturm & Reinhartz-Berger, 2004), which facilitates the instantiation of an application model from a domain model and its validation against the domain model.

According to ADOM, when a domain model is instantiated to an application model, the entities in the resulting application model are classified as instances of the entities in the domain model. Furthermore, the application models may include multiple instances of domain-model entities, as well as additional entities. Hence, an application model can be considered as a refinement of the domain model. The validation of an application model against the relevant domain model employs model matching for verification purposes.

Due to the difficulty of assessing structural similarity with respect to models of different abstraction levels, we seek for refinement equivalence rather than structural similarity.

Refinement equivalence is a situation where a detailed (application) model can be perceived as a refinement of a more abstract (domain) model. To this end, we first need to establish an understanding of the nature of the refinement of models. The chapter discusses several types of refinement operations and indicates their structural characteristics, demonstrated by using the object-process methodology (OPM) as a modeling language. Understanding the consequences of model refinement is the basis for an algorithm that identifies structural equivalence of two models.

The remainder of the chapter is organized as follows. The next section briefly introduces the OPM modeling language and provides an overview of the ADOM approach. The following section discusses different refinement operations and illustrates their outcome in an OPM model. Then we describe a rule-based algorithm for identifying structural equivalence of OPM models in the context of validating an application model against a domain model. Following that, a review of related work is presented, and finally a concluding discussion.

Overview of ADOM and OPM

This section starts with a brief introduction to OPM, then provides an overview of the ADOM approach in general and the ADOM-OPM dialect in particular.

Object-Process Methodology

OPM, whose details are provided in Dori (2002), has been applied for various purposes at different development phases and tasks, such as conceptual requirements modeling (Soffer, Golany, Dori, & Wand, 2001), enterprise resource planning (ERP) system modeling (Soffer, Golany, & Dori, 2003), Web application design (Reinhartz-Berger et al., 2002), real-time systems specification (Peleg & Dori, 1999), algorithm specification (Wenyin & Dori, 1998), and others.

OPM incorporates two equally important classes of entities: objects and processes. While object-oriented methods encapsulate processes in objects, and business-process modeling methods represent activities detached from the objects they affect, OPM unifies the system structure and behavior into a single representation. It uses a single graphic tool, the object process diagram, set, as a single view of all the system aspects, both structural and dynamic. Structure is expressed by objects connected with structural relations, such as characterization (e.g., between an object and its attributes), aggregation (part of), specialization (is-a), and general tagged structural relations (specifying any other relation named by a tag). The behavior of a system is represented by a set of procedural links, which can be classified into three classes of links: enabling links, transformation links, and triggering links. Enabling links (e.g., instrument links) relate an object to a process when the presence of the object is required for the process to occur, but this occurrence does not affect the object state or value. Transformation links (e.g., effect links) relate an object to a process that changes the object state or value (including its creation and destruction). Triggering links (e.g., event links) relate a transformation of an object (reflected in its state or value) to a process it triggers.

Similar to other modeling languages (e.g., DFD), OPM allows the refinement of a model by zooming into processes and unfolding the structure of objects to enable a top-down analysis. The resulting model is a hierarchical OPD set, which specifies all the aspects of a system at a spectrum of detail levels.

A part of OPM notation is given in Figure 1.

Figure 1. OPM notation

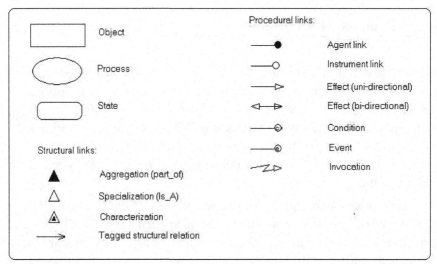

Application-Based Domain Modeling

ADOM is a generic domain analysis approach, facilitating the creation of domain models, their instantiation for creating application models, and the validation of the resulting application models. Being influenced by the classical framework for metamodeling presented in OMG (2006), the ADOM approach is based on a three-layered architecture: application, domain, and language. The application layer, which corresponds to the model layer (M1), consists of models of particular systems, including their structure and behavior. The language layer, which corresponds to the metamodel layer (M2), includes metamodels of modeling languages, such as UML (unified modeling language), OPM, and so forth. The intermediate domain layer consists of domain models. The ADOM approach enforces constraints among the different layers; in particular, the domain layer enforces constraints on the application layer, while the language layer enforces constraints on both the application and domain layers.

Including language metamodels as an upper layer, the ADOM approach is language independent. However, in practice, language-specific ADOM dialects must be used. Such dialects include ADOM-UML (Reinhartz-Berger & Sturm, 2004; Sturm & Reinhartz-Berger, 2004) and ADOM-OPM (Sturm, Dori, & Shehory, 2006), which is the dialect used in this chapter, too.

ADOM-OPM extends OPM with two new features: (a) a multiplicity indicator, which is attached to entities at the domain layer and constrains the number of entities of that kind that can appear in a particular application model in that domain, and (b) a role, which is a stereotype-like element emphasizing additional semantics for an OPM entity. Roles are used within application models, classifying entities as instances of domain-model entities. These two features establish the relationships between domain and application models. When an application model is created, its entities are assigned roles that correspond to the entities of the domain model, and the links among them are bound to preserve the corresponding link structure of the domain model. Additional entities can appear in the application model (without assigned roles) as long as they do not violate the domain constraints.

Validating an application model against the domain model entails checking that (a) the multiplicity constraints, specified by the multiplicity indicators, are not violated, that is, the number of entities in the application model that are classified with a certain role complies with the multiplicity indicator of the domain-model entity, and (b) the link structure of the application model is equivalent to the link structure of the domain model, considering their corresponding entities.

Refinement Equivalence

This section discusses different refinement operations and provides observations that characterize their structural impact in an OPM model in order to establish an in-depth understanding of model refinement in general. It should be noted that for the purposes of model retrieval and validation, matching may address models at different abstraction levels. The retrieval of a complete and detailed model requires its matching against a preliminary or partial input model, which is at a higher abstraction level than the retrieved model. Similarly, the validation of an application model against a domain model requires the matching of a low-level detailed (application) model against a high-level (domain) model. However, model matching as addressed in the literature so far has mainly dealt with models whose abstraction levels are identical. Two common similarity aspects (or measures) that are usually checked are entity similarity and structural similarity. Entity similarity assessment (also called semantic similarity) aims at identifying entities that are semantically similar in the models that are being matched. It may employ mechanisms of various accuracy and complexity levels, ranging from the identification of identical entity name and type (Soffer, Golany, & Dori, 2005), through thesaurus-based affinity measurement (Castano, De Antonellis, Fogini, & Pernici, 1998; Ralyte & Rolland, 2001), to concept hierarchy-based distance measurement (Chen-Burger et al., 2000; Lai et al., 1999). Structural similarity assessment, on the other hand, typically follows the links among the entities in one model and searches for parallels in the other model (Chen-Burger et al.; Massonet & Lamsweerde, 1997; Ralyte & Rolland; Sutcliffe & Maiden, 1998). This is sometimes termed neighboring-entities search. According to these two similarity assessments, two models are considered matching if they include the same entities and the same links to some extent. However, in case the models that should be matched are not at the same level of abstraction, then one cannot expect both models to have the same structure and set of links. Rather, while a high-level model specifies a set of entities and relationships among them, the low-level model includes the same entities (or their instances) along with other entities. Therefore, the model structure might be different, including all the other entities that exist in the detailed model and the links among them.

Since the instantiation of a domain model to an application model is a specific case of refinement, specific implications with respect to the application model validation shall be indicated. The most notable characteristic of this specific case is that entities of the application model are classified as instances of entities in the domain model. Hence, semantic similarity assessment techniques (e.g., Palopoli, Sacca, Terracina, & Ursino, 2003; Ralyte & Rolland, 2001) are not needed for matching these models.

We view an OPD as a directed and labeled graph whose nodes are entities (objects and processes) and edges are both structural and behavioral links among the entities. A refinement operation inserts new nodes and edges into an existing graph. These additional parts may replace existing edges, thus they may form paths between nodes that were directly linked in the original graph.

We shall examine and characterize the results of two types of refinement operations: refinement of structure and refinement of behavior. Specifically, we aim at identifying conditions under which a path can be considered equivalent to a given link.

Definition 1: Let A and B be entities, and let P be a path between A and B. P is equivalent to a link of type l if and only if a link l between A and B can be replaced by P through a refinement operation.

Notation: $P \cong l$.

Refinement of Structure

The paths established when structure is refined can replace both structural and procedural links that originally existed with the entity whose structure is being refined. We shall examine these two categories of links and characterize the path that replaces them in a refined model.

Structural links: When more structural details are revealed, a direct structural link in the abstract model can be replaced by a path including structural links and entities. This is demonstrated in the example shown in Figure 2, in which a characterization link between Warehouse and Number of Locations in the abstract model (a) appears as a path including both specialization and characterization links in the refined model (b). The refinement indicates that only a warehouse in which inventory locations are managed is characterized by the attribute Number of Locations.

In general, a path including a number of structural links can always be abstracted to a specific link type independently of the order in which these links appear.

Definition 2: Let L be a set of link types. $l \in$ L is dominant with respect to L if and only if $P \cong l$ is true for every path P that includes l together with any $r \in$ L.

Notation: $D_L = l$.

Considering the example of Figure 2, it is clear that $D_{\{Specialization, Characterization\}}$ = Characterization as inheritance maintains characteristics along the hierarchy. Another example of this dominance is the attribute Number of Wheels, which characterizes a vehicle as well as a car, which is a specialization of a vehicle.

Observation 1: Let A and B be entities and P be a path from A to B. Let L be the set of link types included in P. If $D_L = l$, then $P \cong l$.

Figure 2. Example of refinement involving a structural link

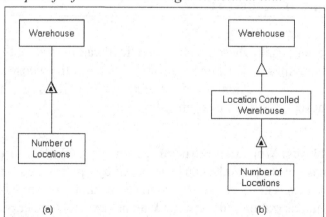

Observation 1 is a direct result of the definition of dominance with respect to a set of link types. It is useful for identifying equivalence regarding paths that include structural links since dominance can easily be established considering these link types, as in the above example. As another example of establishing dominance, consider the attribute Power that characterizes Engine. It characterizes the engine as well as the car of which the engine is part. Hence, characterization is dominant with respect to aggregation as well.

Procedural links: When a procedural link exists between an entity whose structure is being refined and another entity, the resulting path in the refined model consists of both structural and procedural links. As an example, Figure 3a shows an abstract model including an effect link between Engineering Change Processing and Item Technical Data. A refined model (Figure 3b) shows that Item Technical Data is composed of Bill of Material and Routing, which are affected by Engineering Change Processing. A third part of Item Technical Data, Technical Specification, remains intact as it is not even connected to the process.

In general, a refined model may specify the interaction of a process with attributes, parts, or specializations of an entity, whereas an abstract model simply specifies an interaction with the entity.

Figure 3. Example of a procedural link in structure refinement

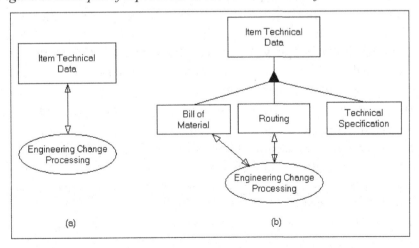

Observation 2: Let A, B, and C be entities. Let P be a path from A to B so that A is linked to C and C is linked to B by a procedural link of type *l*. If the link from A to C is (Characterization) ∨ (Aggregation) ∨ (Specialization), then P ≅ *l*.

The proof of Observation 2 is by a simple demonstration that such refinement is possible (e.g., Figure 3). Note that Observation 2 does not imply the dominance of procedural links with respect to structural links since there may be paths that cannot be abstracted to a procedural link. For example, in Figure 3b, the path between Engineering Change Processing and Technical Specifications is not equivalent to the effect link included in it.

Refinement of Behavior

In general, the refinement of behavior is more difficult to identify than the refinement of structure for reasons that are explained below. Nevertheless, this difficulty is partly overcome when dealing with ADOM's classified entities. We shall first address the general case of refinement when no entity classification is used, and then explain how it becomes easier when ADOM-related models are addressed.

The behavior of a system or a domain is captured by processes. A process can be refined into a sequence of activities (subprocesses) that comprise it. Such a sequence is modeled as a path leading from an initial state (or input objects) to a final state (or output objects). The subprocesses in a refined process may interact with other objects besides the ones the higher level process interacts with, but these objects can be considered internal, meaning that in the abstract view of the process, the interaction is not observed. For example, consider two people who perform a task together. The interaction and allocation of work between them is internal in the sense that it is not of interest to others as long as the job is done.

The difficulty in identifying a refined process lies in the fact that unlike the refinement of structure, in which a link is replaced by a path, when a process is refined, an entity is replaced by a path (or several paths). Therefore, the initial and final states are the only reference points available. However, this information is not always sufficient for a conclusive identification of refinement equivalence. Consider a process of a high level of abstraction (e.g., building a house), having an initial state (existing plans, resources) and a final state (a house built). This process can be refined into many different processes, all having the same initial and final states and subset of interactions (stakeholders, authorities, building materials) as the abstract one. Yet, while being all equivalent to the abstract model, these refined processes are not equivalent to one another. As a detailed example, consider the abstract process of Supplying Customer Order in Figure 4a, which can be refined into the two different processes in Figure 4b and c. These two refined processes have identical initial and final states, Open Customer Order and Delivered Customer Order, respectively, as does the abstract process. However, while

Figure 4. An abstract model and two possible refinements

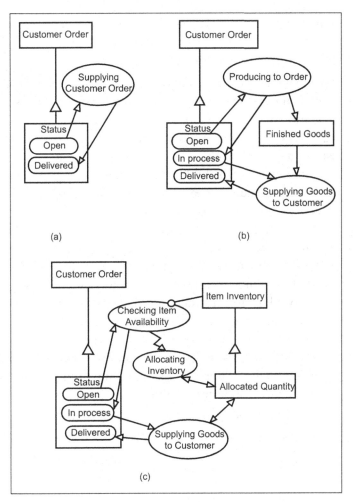

both processes can be considered equivalent to the abstract model, they are not equivalent to one another (in their internal division into subprocesses, additional inputs and outputs, etc.). It is therefore easier to formulate a necessary condition rather than a necessary and sufficient condition for refinement equivalence of processes.

Observation 3: Let m1 be a model portion in which process A transforms an initial state s_1 into a final state s_2. Let E1 be the set of entities directly linked to A in m1. Let m2 be a model portion that refines m1. Then m2 consists of a path P and a set E2 of entities that are directly linked to the entities of P so that P is from an initial state s_1 to a final state s_2 and E1 $\subseteq$ E2.

Note that the initial and final states are not necessarily explicitly represented in an abstract model, in which case the inputs and outputs of the process should be considered in a similar manner to the states.

Observation 3 provides a necessary condition that might not be sufficient for the identification of equivalence. When the lower level model is a result of an instantiation operation of a domain model, its entities are assigned roles that correspond to domain-model entities. In other cases, we need a way to relate the subprocesses in a refined model to a process in the abstract model. For that purpose, we note that it is likely that at least one of the subprocesses in a refined model bears a name that can be identified as similar to the general process' name as appears in the abstract model. Such resemblance can be detected by existing affinity detection techniques, which are not the focus of this chapter. This can be explained by a tendency to name the process in the abstract model after the main activity that constitutes the essence of the process. In fact, such tendency is not unique to process models. Suggesting a semiautomatic procedure for abstracting a database schema, Castano et al. (1998) refer to a "representative" element of the detailed schema, whose name should be given to the generalizing element in the abstracted schema. When refining an abstract process to lower abstraction levels, details of other activities are revealed. In the example of Figure 4, Supplying Goods to Customer can be identified as similar to Supplying Customer Order.

In such cases, we expect the refined model to include a path from the initial state to the similarly named process (or, in ADOM-based models, to the process whose role corresponds to the process in the domain model) and to the final state. A path is also expected to relate the process to other entities that interact with it in the higher-abstraction-level model. If such paths exist in a detailed model, and if they are equivalent to the links of the abstract model, than the detailed model can be considered as a refinement of the abstract one. Observation 4 indicates a condition under which a path that may include a number of processes and objects or states is considered as equivalent to a specific type of procedural link.

Observation 4: Let A be an object or a state of an object, B be a process, and P be a path between A and B. Let l be the procedural link by which A is related to P, then $P \cong l$.

Note that the direction of the path can be from the object to the process or backward, depending on the specific links involved.

Observation 4 can be justified when abstracting the entire path (processes and objects) to a process (named after its representative activity, B). The link that determines the nature of the interaction between this abstracted process and the object is the link relating the object to the path. In the example of Figure 4b and c, the path from the state Open of Customer Order Status to Supplying Goods to Customer is equivalent to the direct link from Open to Supplying Customer Order in 4a.

Observation 4 provides a sufficient condition for identifying refinement equivalence. However, this condition, though sufficient, is not a necessary one. It is based on the assumption, discussed above, that the abstract process is named after its main activity. This assumption is not necessarily always true. For example, a production process can be refined into processes of cutting, drilling, milling, and so forth. In such cases, the path between the initial and final states in the abstract model has to be matched against the path in the detailed model. That path can be decomposed into individual links for this purpose. As explained above, when application-model processes bear roles that classify them as corresponding to domain-model processes, the naming difficulty does not exist. Thus, Observation 4 can conclusively identify refinement equivalence.

Tracking Refinement Equivalence

The previous section identified conditions that enable the detection of refinement equivalence. When an application model is validated against a domain model, the following steps can be taken: (a) The names of the entities that have a role assigned to them in the application model are replaced by their roles, (b) satisfaction of the multiplicity constraints specified in the domain model is determined, and (c) the links among the entities in the domain model are matched by corresponding links in the application model. In case such corresponding link is not found, an equivalent path is searched for between the source entity and the destination entity of the link.

This section describes a rule-based algorithm that identifies refinement-equivalent paths with respect to a given link type. The algorithm is basically a path-searching algorithm applying rules, which follow the discussion and observations of the previous section, to assure that the path found is indeed equivalent to the link being matched.

Searching for an Equivalent Path

Consider a pair of OPDs <A, D>, where A is the application model and D is the domain model being matched. Assume A is searched for a path between two entities that are directly related in D. The steps of the search shall first be informally described, and then specified formally. Each step of the search partitions A into two sets of entities: One is the set of entities to which a path from the source entity is already established, and the other is the set of entities that are not yet explored. Starting from the source entity, each step follows a link and moves one entity from the unexplored set to the set of entities that are connected to the source. The choice of link to be followed is based on the search rules, whose details are given below. The steps repeat until a direct link is found from the connected set of entities to the destination entity, or until all the links have been exhausted and it is clear that the searched-for path does not exist. The algorithm seeks to establish the existence of a path that is not necessarily the shortest path, hence no backtracking is performed and the number of steps is at most the number of entities in A minus one.

The formal specification of the search applies to the following notation:

s: the source entity of the link in D whose equivalent path is being searched for in A.

d: the destination entity of the link in D whose equivalent path is being searched for in A

- $L_M(e_1, e_2)$: Let e_1 and e_2 be entities; then $L_M(e_1, e_2)$ is a Boolean variable whose TRUE value indicates the existence of a direct link from e_1 to e_2 in model M (M is either the application model A or the domain model D).
- $Link_M(S_1, S_2)$: Let S_1 and S_2 be nonoverlapping sets of entities in model M; then $Link_M(S_1, S_2)$ is an indicator expressing the existence of a direct link from an entity in S_1 to an entity in S_2.

$$Link_M(S_1, S_2) = \begin{cases} 1 & \text{if } \exists\ e_1 \in S_1, e_2 \in S_2, \text{ such that } L_M(e_1, e_2) = \text{TRUE} \\ 0 & \text{otherwise} \end{cases}$$

- S_M: the set of entities in model M
- $C_i(M, s)$: the set of entities in model M to which a path from s has been found until the ith step of the search
- $U_i(M, s)$: The set of entities in model M whose relationship with s has not yet been investigated by the ith step of the search

In the context of the application model, $C_i(A, s)$ and $U_i(A, s)$ partition S_A so that at each step i of the search, $S_A = C_i(A, s) + U_i(A, s) + \{d\}$. In other words, each entity in A belongs either to the set of entities that have already been established as linked to s (including s itself) or to the set of entities whose relationship with s is unknown yet, or to the set that holds d only.

Lemma: Let an application model A be searched for a path from s to d at the ith step of the search. A path from s to d exists only if Max [$Link_A (C_i(A, s), \{d\})$, $Link_A (C_i(A, s), U_i(A, s))*Link_A (U_i(A, s), \{d\})$] = 1.

Proof: Assume a path exists. It can lead from $C_i(A, s)$ directly to d, then $Link_A(C_i(A, s), \{d\}) = 1$. Otherwise, it leads from $C_i(A, s)$ to some entity

$e \in U_i(A, s)$ and from e to d. Then $\text{Link}_A(C_i(A, s), U_i(A, s)) = 1$ and $\text{Link}_A(U_i(A, s), \{d\}) = 1$.

Assume a path does not exist. Then $\text{Link}_A(C_i(A,s), \{d\}) = 0$ and the following are true:

1.   If $\text{Link}_A(C_i(A, s), U_i(A, s)) = 1$, then $\text{Link}_A(U_i(A, s), \{d\}) = 0$.
2.   If $\text{Link}_A(U_i(A, s), \{d\}) = 1$, then $\text{Link}_A(C_i(A, s), U_i(A, s)) = 0$.

Note that the above lemma is one sided; that is, it does not imply that if Max $[\text{Link}_A(C_i(A, s), \{d\}), \text{Link}_A(C_i(A, s), U_i(A, s))*\text{Link}_A(U_i(A, s), \{d\})] = 1$, then a path exists. Rather, this is a necessary condition for the existence of such a path.

The initial state of the search is $C_0(A, s) = s$, $U_0(A, s) = S_A - \{s, d\}$. At each step, if the condition specified in the lemma is satisfied, one entity is moved from $U_i(A, s)$ to $C_i(A, s)$ by following a link, implying that a relation of this entity to s is established. The steps repeat until either a path is found, that is, $\text{Link}_A(C_i(A, s), \{d\}) = 1$, or the condition of the lemma is not satisfied; that is, the searched-for path does not exist. The search rules ensure that the found path is equivalent to the link being searched for.

Figure 5 specifies the equivalence path search algorithm. This algorithm employs the following operations.

Figure 5. Equivalent path search algorithm

```
Current = s
Fold_Structure (d)
Exclude_Links
Do while (LinkA(Ci(A, s),Ui(A, s))*LinkA(Ui(A, s),{d}) = 1)
  AND (LinkA(Ci(A, s),{d}) <> 1)
      If Link_Type is procedural then Fold_Structure(Current)
      Exclude_Links
      Verify_Equivalence
      If Link_Type is structural then Compute_Cardinality
      Select_Entity
End Do
If    (LinkA(Ci(A, s),{d}) = 1) AND (Path_Cardinality =
      Link_Cardinality) AND (Condition) then Path_Found =
      TRUE
Else  Path_Found = FALSE
```

Fold_Structure (entity): A folding operation of structural relations in OPM is an abstraction operation in which a detailed OPD portion, including structural relations such as characterization, aggregation, and specialization, is replaced by an OPD portion of a higher abstraction level. The entities that provide the structure details of the entity being folded (which is the parameter of this operation) are not shown in the abstracted OPD. Other entities, which are originally related to the structure details, are related directly to the folded entity.

This operation is employed only when the link, whose equivalent path is searched for, is a procedural link. Its role is to replace paths created through refinement of structure by their equivalent procedural links on the basis of Observation 2.

Exclude_Links: This operation excludes links that cannot be included in the path. Links can be excluded from the search for three reasons. The first reason is that they cannot be part of the path according to the search rules, in which case they are excluded at the beginning of the search. The second reason is that their direction is opposite of the search direction. At every step of the search, the unidirectional links from the entities of $U_i(A, s)$ to the entities of $C_i(A, s)$ are excluded from the search. The last reason applies to inheritance (is-a) links, which may be included in a path in both directions, from the special to the general as well as the other way. When going up the relation, the links to other specializations of the general entity cannot be included in the path.

Select_Entity: At every step of the search, all the links from the entities of $C_i(A, s)$ to the entities of $U_i(A, s)$ are arranged according to priorities defined by the search rules. The first link according to this order is selected and the entity it relates to is moved to $C_i(A, s)$ and becomes the Current entity.

Verify_Equivalence: The search rules specify for a given link the link type that must be included in the path and its required position (at the source, at the destination, or anywhere in the path). If the required position is at the source or destination of the path, then all the links from s or to d (respectively), which are not of the mandatory type (i.e., are not of the type that must be in that position in the path in order to preserve the nature of the interaction), are excluded from the search at the first step by the Exclude_Links operation.

As a result, a Boolean variable Condition is assigned a TRUE value. If the required position is anywhere in the path, the Condition is verified by a set of indicators EC_e, defined next.

Let e be an entity in $C_i(A, s)$; then $EC_e = 1$ if and only if a link of the mandatory type is in the path from s to e.

Starting at $EC_s = 0$, and letting e be moved from $U_i(A, s)$ to $C_i(A, s)$ through a link of type t from an entity $a \in C_i(A, s)$, then:

$$EC_e = \begin{cases} 1 & \text{if } (EC_a = 1) \text{ or } (t \text{ is of mandatory type}) \\ 0 & \text{otherwise} \end{cases}$$

When a path is found, $EC_d = 1$ implies that it includes at least one link of the mandatory type (according to the conditions specified by the search rules), in which case Condition = TRUE.

Compute_Cardinality: This operation is performed only when structural relations are searched for. The cardinality of a link is defined as <SL, SU, DL, DU>, where SL is the source lower participation constraint, SU is the source upper participation constraint, DL is the destination lower participation constraint, and DU is the destination upper participation constraint.

Let e be an entity in $C_i(A, s)$; then the aggregated cardinality of the path from s to e is denoted by $<SL_e, SU_e, DL_e, DU_e>$, where s holds <1, 1, 1, 1>.

Let a be moved to $C_i(A, s)$ through a link whose cardinality is <SL, SU, DL, DU> from entity $e \in C_i(A, s)$, then $SL_a = SL_e * SL$, $SU_a = SU_e * SU$, $DL_a = DL_e * DL$, $DU_a = DU_e * DU$.

For example, assume an item is supplied by zero to three suppliers, a supplier has one to two contact persons, and a supplier can supply one or more (1...m) items. The aggregated cardinality of the path between an item and a purchasing contact person is <1, m, 0, 6>.

Search Rules

The search for an equivalent path employs rules of two types: link selection rules and equivalence conditions. Both rule types are defined for each type of link in OPM. A link selection rule defines the types of links that can be included in an equivalent path and provides searching priorities for the search algorithm. It is applied by the Exclude_Links operation, which excludes all the irrelevant links from the search, and by the Select_Entity operation, which uses the priorities given for selecting the entity to be moved from $U_i(A, s)$ to $C_i(A, s)$. An equivalence condition defines conditions for a path to be equivalent to a link of a certain type. It is employed by the Verify_Equivalence and Exclude_Links operations. Conditions may specify link types that must be included in a path and their required positions that can be at the source of the path, at its destination, or at any point in the path.

A link selection rule is of the following form:

Link Selection (Link Type): {Set of Types}

Link Type is the type of link to which the path is to be equivalent, while Set of Types is an ordered set of link types. All the link types in the set can be included in a path, which is equivalent to Link Type. Their order in the set determines the priority in which the search algorithm considers links in the examined OPD when searching for a path.

On the basis of Observation 1, the Set of Types specified for structural link types satisfies $D_S = l$, where l is the Link Type and S is the Set of Types.

For example, the link selection rule for aggregation, which is a structural link that denotes a whole-part relation and is dominant with respect to specialization (is-a) relations only, is:

Link Selection (Aggregation): {Aggregation, Specialization}

For procedural link types, the Set of Types is defined on the basis of Observation 4. According to this observation, the link that determines the equivalence is the one related to the source or destination object without restrictions on the types of links in the path. Hence, the Set of Types for procedural link types includes all the types of links in OPM.

The order of the types in the Set of Types always sets the relevant Link Type as the first priority for the search algorithm. For procedural link types, it lets the algorithm prefer procedural links over structural ones.

An equivalence condition is of the following form:

Equivalence Condition (Link Type): Mandatory Type must be located at Required Position in the path

Mandatory Type is a link type that is necessarily included in the path in order to preserve the nature of the interaction, where Required Position is the exact position where it should appear (the possible values are at Source Position, at Destination Position, and Anywhere).

Mandatory Type is, with one exception, the Link Type itself. The exception is an invocation link, which represents the triggering of a process by the completion of another process. This can also be modeled as an event created by the first process and triggering the second one. In this case, an event link replaces the invocation link.

For structural link types, the Required Position is Anywhere, since the link selection rules ensure the dominance of the specific link type with respect to the links in the path. Hence, their position in the path is of no importance as long as they are present. For procedural link types, the Required Position, according to Observation 4, depends on the link type. Links whose direction is from the object to the process (e.g., instrument links) require the Mandatory Type at the source of the path, while links that lead from the process to the object (e.g., result links, which are unidirectional effect links) require the Mandatory Type at the destination of the path.

For example, below are the equivalence conditions for aggregation links (i.e., structural links that denote whole-part relations) and instrument links (i.e., procedural links that denote input objects that are not changed by the process; these links are directed from the object to the process).

Equivalence Condition (Aggregation): Aggregation must be located Anywhere in the path

Equivalence Condition (Instrument Link): Instrument Link must be located at Source Position in the path

As explained above, the two types of rules are based on Observation 1, which addresses structural links when structure is refined, and on Observation 4, which addresses procedural links when behavior is refined. Observation 2, which addresses procedural links when structure is refined, is not applied as part of the rule base, but is taken into account by the Fold_Structure operation performed by the search algorithm.

Exemplifying the Equivalent-Path Search Algorithm

The algorithm steps are illustrated by an example given in Figure 6: Figure 6a is part of a domain model, while Figure 6b is an application model that should be matched against the domain model. The domain model specifies the main concepts as well as their multiplicity constraints. For example, Pro-

Figure 6. Refinement equivalence example

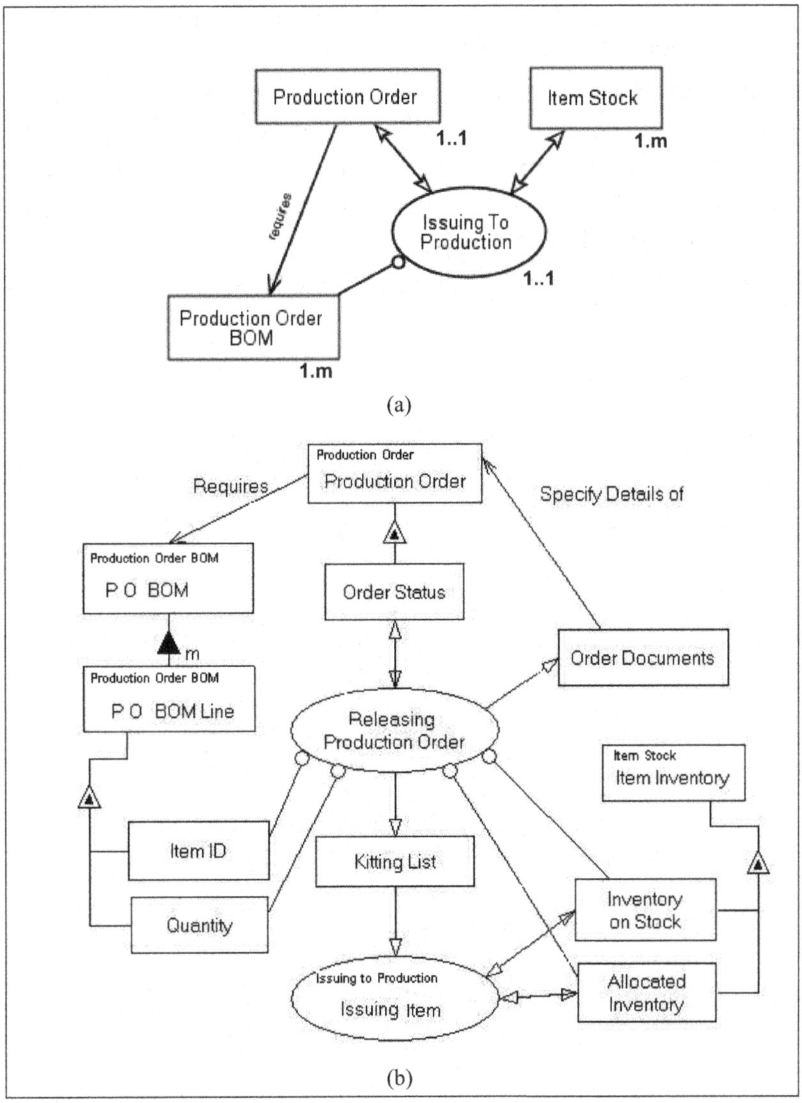

duction Order and Issuing to Production are indicated as mandatory single entities (the 1..1 at the right lower corner of the entities), meaning they must be instantiated exactly once in any application model of the domain, while Production Order BOM and Item Stock are indicated as mandatory multiple entities (the 1..m at the right lower corner of the entities), meaning they must appear at least once in any application model in the domain. Correspondingly, some of the application-model entities have roles (at their left upper corner) that relate them to the domain-model entities, while others are additional application-specific entities. Note that the number of role-classified entities in the application model is consistent with the multiplicity indicators specified in the domain model for each role.

None of the procedural links specified in Figure 6a appears as a direct link in Figure 6b. Nevertheless, they are all matched by equivalent paths in the application model. The domain model specifies that a process of Issuing to Production affects the Production Order and the Item Stock, and uses the Production Order BOM (which specifies the required materials). In the application model, a process of Releasing Production Order precedes Issuing

Figure 7. Search algorithm 1st step

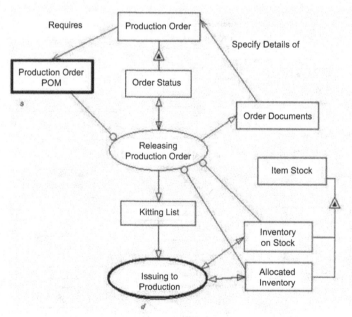

Item (whose role is Issuing to Production), using Item Inventory (whose role is Item Stock) information as well as the item ID and quantities specified by PO BOM Lines, which are parts of the PO BOM (both have a role of Production Order BOM). The process of Releasing Production Order creates Order Documents (a set of documents, specifying details of the production order, to be used in the production process) and a Kitting List, which is a list of items to be prepared in kits before they can be issued to production. The Issuing Process uses the Kitting List and affects the Item Inventory.

We shall follow the steps of the search algorithm for tracking an equivalent path that matches the instrument link from Production Order BOM to Issuing to Production in the domain model of Figure 6(a) in the application model of Figure 6(b). Two entities in that model are classified with the role of Production Order BOM. However, since one is part of the other, we will use the whole as the source of the searched path, as illustrated in Figures 7 to 10. The search in Figures 7 to 10 is performed after the names of the entities have been replaced by their roles (whenever they have one), according to the first validation step.

Figure 8. Search algorithm 2nd step

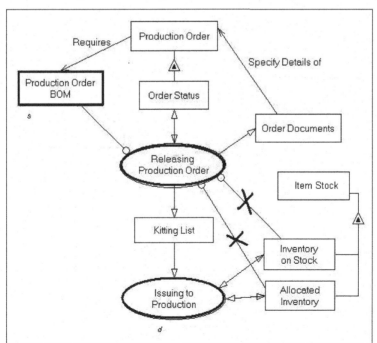

Step 1 (see Figure 7): $C_0(A, s)$ includes the source entity, Production Order BOM (highlighted). The source entity is the Current entity, and a Fold_Structure(Current) operation is performed. As a result, its structural details are not seen, and the instrument links originally related to these details are now related directly to Production Order BOM itself. $U_0(A, s)$ includes all the other entities in the model, except the source entity, Production Order BOM, and the destination entity, Issuing to Production (highlighted). $C_0(A, s)$ is linked to $U_0(A, s)$, which is linked to the destination entity, thus the condition of the lemma is satisfied.

Step 2 (see Figure 8): Following the instrument link, $C_1(A, s)$ includes Releasing Production Order in addition to Production Order BOM. Note that the equivalence condition of an instrument link requires that the first move should be through an instrument link, and it is satisfied. Two instrument links that lead to Releasing Production Order are excluded from the search by the Exclude_Links operation since their direction is opposite of the path direction. $C_1(A, s)$ is still linked to $U_1(A, s)$, which is linked to the destination entity.

Figure 9. Search algorithm 3ʳᵈ step

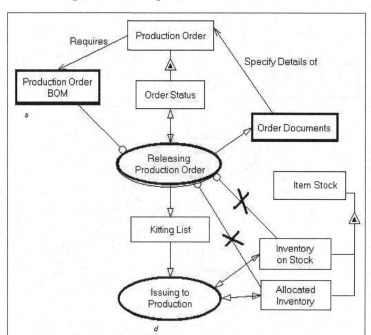

Step 3 (see Figure 9): Following the effect link, Order Documents is included in $C_2(A, s)$. Note that this is a random choice from the three effect links that lead from Releasing Production Order. $C_2(A, s)$ is still linked to $U_2(A, s)$, which is linked to the destination entity.

Step 4 (see Figure 10): Following the next effect link from $C_2(A, s)$, Kitting List is now added to $C_3(A, s)$. $C_3(A, s)$ is now linked to the destination entity, thus establishing a path that meets the equivalence conditions, and is therefore equivalent to the direct link of the domain model.

Note that Step 3 is actually redundant and could be avoided by a different choice of link. Nevertheless, by addressing all the links of the $C_i(A, s)$ set, the algorithm is able to simply look one step ahead at a time and avoid a recursive backtracking.

The complexity of the search algorithm is $O(|S_A|)$, where $|S_A|$ is the number of entities in A. The search is performed for each link in D when the models

Figure 10. Search algorithm 4ᵗʰ step

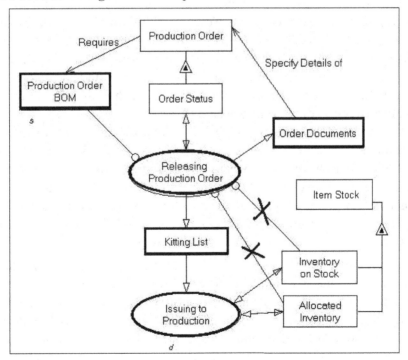

are matched. Hence, the complexity of the matching is $O(|S_D|^2 * |S_A|)$. Note that $|S_D|$ is expected to be significantly smaller than $|S_A|$.

Related Work

Model similarity has been addressed by several disciplines. The ones that are relevant to this work are the disciplines of reuse and schema analysis and matching. The difference in abstraction level between matched models has not, to the best of our knowledge, been explicitly addressed in the reuse literature. Kim (2001) presents an object-oriented model reuse application in which an initial model, including classes and nonspecific links, serves as a basis for retrieving an existing complete model. The retrieved model is then modified and adapted to the current needs using modification rules, whose details are not presented. No details are available about how a complete model is retrieved and evaluated, how this retrieval considers the nonspecific links of the input model, and how structurally different from each other the models retrieved are.

Structural similarity plays an important role in the works that deal with analogical reasoning (Massonet & Lamsweerde, 1997; Sutcliffe & Maiden, 1998), where models designed for a certain domain are applied to other domains by analogy. The retrieval is based on structural properties of the model and on semantics, which is based on generalizations. In Sutcliffe and Maiden, the models to be retrieved include a number of layers, each dealing with different information types, going from an abstract layer to a detailed one. The matching with the input information interactively follows these layers of specific information types, and the user is required by the system to provide enough information to discriminate between existing models. Hence, the structural similarity deals with models of the same abstraction level. In Massonet and Lamsweerde, while the entities of an input model are generalized to a higher level in an is-a hierarchy, their link structure is expected to remain the same and serves as a basis for structural similarity assessment.

Other works that apply reuse for method engineering (Ralyte & Rolland, 2001) and for enterprise modeling (Chen-Burger et al., 2000) use simple structural similarity assessment along with semantic similarity based on affinity (Ralyte & Rolland) or on a generalization hierarchy (Chen-Burger et al.). The model used by Ralyte and Rolland includes multiple abstraction levels. Hence, there might be a match between the abstraction level of a query model and one of the levels of the reusable models, but it is not explicitly addressed and verified. None of the above reviewed works relates to model matching for validation purposes as proposed in this chapter.

Schema-matching literature focuses on the semantic mapping of one schema to the other. While semantic similarity in the reuse literature is mostly affinity based, or in some cases relies on is-a hierarchies, semantic matching in the schema-matching literature sometimes combines the affinity of terms with structural considerations. Schema matching maps elements of one schema to elements of another schema rather than compute similarity measures between the two schemas. Hence, each pair of elements is thoroughly examined and structural aspects, such as attributes and is-a relations, are taken into account (Madhavan, Bernstein, & Rahm, 2001; Rahm & Bernstein, 2001; Rodriguez & Egenhofer, 1999). In some cases, paths are sought where direct links do not exist (Palopoli et al., 2003). Nevertheless, dealing with schemas means dealing with a low level of abstraction. Some schemas may be more detailed than others, and the techniques suggested are aimed at overcoming such differences rather than at dealing with models that are basically at different abstraction levels. Typical to this situation is the use of the term "structural equivalence" of schemas (Algaic & Bernstein, 2001), which relates to a consistent mapping of schema elements from one schema to another and backward in the lowest abstraction level. It is defined as structural as opposed to semantic equivalence, which relates to integrity constraints as well.

The similarity assessment of entities, presented by Rodriguez and Egenhofer (1999), relates to parts, functions, and attributes of two entity classes. Their similarity measure uses a function that provides asymmetric values for entity classes that belong to different levels of abstraction. While addressing single entity classes, they take contextual information into account for the similarity measurement. However, context information of an entity cannot be considered equivalent to a view of the entity as a part of a model, including relationships with other entities.

A more holistic view of schema analysis, including a variety of techniques for schema abstraction, matching, and reuse, is presented in Castano et al. (1998). Schema abstraction is an operation in the opposite direction compared to our discussion of refinement operations. The ERD schemas addressed limit the discussion to structural links only, without addressing the representation of behavior. Yet, their abstraction operation is consistent with our opposite-direction refinement, and applying the algorithm presented here to their examples of detailed and abstract schema yields a match. A number of schema similarity measures are presented there, dealing mainly with semantics and, to a limited extent, model structure, particularly with attributes. Interestingly enough, their fuzzy similarity measure is asymmetric and may indicate that schema a matches schema b to a higher extent than in the other direction. This is explained as being a result of differences in the abstraction level between the two schemas.

Our approach can be classified according to the extensive classification of schema-matching approaches presented by Shvaiko and Euzenat (2005). It is a structure-level approach (computes mapping elements by analyzing how entities appear together in a structure), syntactic (interprets the input in function of its sole structure following some clearly stated algorithm), and graph based (addressing children, leaves, and relations of entities). However, this classification does not relate to differences in the abstraction level of the matched schemas, and this issue is not addressed by any of the works surveyed there.

In summary, the main contribution of this chapter as compared to related earlier model-matching works is in explicitly addressing models of different abstraction levels, representing both the structure and behavior of a domain of applications.

Conclusion

The reuse of models requires activities that in many cases employ model matching. In this chapter, we stressed that differences in the abstraction level are likely to exist between models, specifically in the retrieval and validation activities, and therefore refinement equivalence is a better measure than structural similarity. Refinement equivalence is identified when a detailed model can be considered a refinement of a model of a higher abstraction level. In this chapter we discussed the notion of refinement equivalence as an enabler of validating a detailed application model against an abstract domain model in the context of the ADOM approach for domain analysis.

The discussion of refinement operations and the observations that characterize their impact on model structure, as well as the search algorithm, address OPM models. However, ADOM is language independent and can be used with other modeling languages as well. Other modeling languages are different mainly in the separation of structural and behavioral aspects of the modeled domain (and applications). Yet, the notion of refinement equivalence is of relevance to models independently of the modeling language. Some of the observations made in this chapter can easily be generalized and become applicable to other modeling languages. For example, Observation 1, which deals with the dominance of structural relations in a path, is not specific to OPM only. Hence, when dealing with models that capture structural information only (e.g., ERD, UML class diagrams), the algorithm can be applied using the search rules that relate to structural links only, omitting the Fold_Structure operation. Regarding the behavioral aspects, generalization is less straightforward. In multiview modeling languages, such as UML, consistency among views might also need consideration.

An equivalent-path search algorithm is, naturally, language specific, and apparently needs to be developed for each modeling language. However, the algorithm presented here is mainly a path-searching algorithm, while specific features of the OPM links are captured by the equivalence rules. Hence, the main body of the algorithm might be applicable to other modeling languages while the unique features of the language might affect mainly the equivalence rules.

The search algorithm that enables refinement-equivalence identification has been implemented in a reuse application that supports business-process alignment and gap analysis in the implementation of ERP systems (Soffer et al., 2005). The application matches abstract enterprise requirement models with a detailed model of the ERP system, and retrieves the parts that match the requirements.

Future research should extend the refinement-equivalence concept and apply it to other modeling languages that serve in reuse applications, such as UML.

References

Algaic, S., & Bernstein, P. A. (2001). A model theory for generic schema management. In *Proceedings of DBPL* (LNCS 2397, pp. 228-246). Berlin, Germany: Springer-Verlag.

Castano, S., De Antonellis, V., Fogini, M. G., & Pernici, B. (1998). Conceptual schema analysis: Techniques and applications. *ACM Transactions on Database System, 23*(3), 286-333.

Chen-Burger, Y. H., Robertson, D., & Stader, J. (2000). A case-based reasoning framework for enterprise model building, sharing and reusing. In *Proceedings of the ECAI Knowledge Management and Organization Memories Workshop*, Berlin, Germany.

Dori, D. (2002). *Object process methodology: A holistic systems paradigm.* Heidelberg, Germany: Springer Verlag.

Eckstein, S., Ahlbrecht, P., & Neumann, K. (2001). Increasing reusability in information systems development by applying generic methods. In *Advanced information systems engineering* (LNCS 2068, pp. 251-266). Berlin, Germany: Springer-Verlag.

Kim, Y. J. (2001). An implementation and design of COMOR system for OOM reuse. In *Active Media Technology, 6th International Computer Science Conference* (LNCS 2252, pp. 314-320). Berlin, Germany: Springer-Verlag.

Lai, L. F., Lee, J., & Yang, S. J. (1999). Fuzzy logic as a basis for reusing task-based specifications. *International Journal of Intelligent Systems, 14*(4), 331-357.

Madhavan, J., Bernstein, P. A., & Rahm, E. (2001). Generic schema marching with Cupid. In *Proceedings of the VLDB Conference*, Rome.

Massonet, P., & Lamsweerde, A. V. (1997). Analogical reuse of requirements frameworks. In *Proceedings of the Third IEEE Symposium on Requirements Engineering (RE'97)* (pp. 26-37).

Mili, H., Mili, F., & Mili, A. (1995). Reusing software: Issues and research directions. *IEEE Transactions on Software Engineering, 21*(6), 528-561.

OMG. (2006). *Meta-object facility (MOF™), version 2.0.*

Palopoli, L., Sacca, D., Terracina, G., & Ursino, D. (2003). Uniform techniques for deriving similarities of objects and subschemes in heterogeneous databases. *IEEE Transactions on Knowledge and Data Engineering, 15*(2), 271-294.

Peleg, M., & Dori, D. (1999). Extending the object-process methodology to handle real time systems. *Journal of Object Oriented Programming, 11*(8), 53-58.

Pernici, B., Mecella, M., & Batini, C. (2000). Conceptual modeling and software components reuse: Towards the unification. In *Information systems engineering: State of the art and research themes* (pp. 209-220). London: Springer-Verlag.

Rahm, E., & Bernstein, P. A. (2001). A survey of approaches to automatic schema matching. *The VLDB Journal, 10*(4), 334-350.

Ralyte, J., & Rolland, C. (2001). An assembly process model for method engineering. In *Advanced information systems engineering* (LNCS 2068, pp. 267-283). Berlin, Germany: Springer-Verlag.

Reinhartz-Berger, I., Dori, D., & Katz, S. (2002). Open reuse of component designs in OPM/Web. In *Proceedings of the 26ᵗʰ Annual International Computer Software and Applications* (pp. 19-24).

Reinhartz-Berger, I., & Sturm, A. (2004). Behavioral domain analysis: The application-based domain modeling approach. In *Proceedings of the 7ᵗʰ International Conference on the Unified Modeling Language (UML2004)* (LNCS 3273, pp. 410-424). Berlin, Germany: Springer-Verlag.

Rodriguez, M. A., & Egenhofer, M. J. (1999). Putting similarity assessments into context: Matching functions with the user's intended operations. In *Proceedings of CONTEXT'99* (LNAI 1688, pp. 310-323). Berlin, Germany: Springer-Verlag.

Shvaiko, P., & Euzenat, J. (2005). A survey of schema-based matching approaches. *Journal on Data Semantics, 4*, 146-171.

Soffer, P. (2005). Refinement equivalence in model-based reuse: Overcoming differences in abstraction level. *Journal of Database Management, 16*(3), 21-39.

Soffer, P., Golany, B., & Dori, D. (2003). ERP modeling: A comprehensive approach. *Information Systems, 28*(6), 673-690.

Soffer, P., Golany, B., & Dori, D. (2005). Aligning an ERP system with enterprise requirements: An object-process based approach. *Computers in Industry, 56*(6), 639-662.

Soffer, P., Golany, B., Dori, D., & Wand, Y. (2001). Modelling off-the-shelf information systems requirements: An ontological approach. *Requirements Engineering, 6*(3), 183-198.

Sturm, A., Dori, D., & Shehory, O. (2006), Domain modeling with object-process methodology. In *Proceedings of the Eighth International Conference on Enterprise Information Systems*.

Sturm, A., & Reinhartz-Berger, I. (2004). Applying the application-based domain modeling approach to UML structural views. In *Proceedings of the 23rd International Conference on Conceptual Modeling (ER2004)* (LNCS 3288, pp. 766-779). Berlin, Germany: Springer-Verlag.

Sutcliffe, A., & Maiden, N. A. (1998). The domain theory for requirements engineering. *IEEE Transactions on Software Engineering, 24*(3), 174-196.

Wenyin, L., & Dori, D. (1998). Object-process diagrams as an explicit algorithm specification tool. *Journal of Object-Oriented Programming, 12*(2), 52-59.

Zhang, Z., & Lyytinen, K. (2001). A framework for component reuse in a meta-modelling-based software development. *Requirements Engineering, 6*(2), 116-131.

Chapter V

On the Use of Object-Role Modeling for Modeling Active Domains

Patrick van Bommel,
Radboud University Nijmegen, The Netherlands

Stijn Hoppenbrouwers,
Radboud University Nijmegen, The Netherlands

Erik Proper,
Radboud University Nijmegen, The Netherlands

Theo van der Weide,
Radboud University Nijmegen, The Netherlands

Abstract

This chapter is about how the object-role modeling (ORM) language and approach can be used for the integration, at a deep and formal level, of various domain modeling representations and viewpoints, with a focus on the modeling of active domains. The authors argue that ORM is particularly suited for enabling such integration because of its generic conceptual nature; its useful, existing connection with natural language and controlled languages; and its formal rigor. They propose the logbook paradigm as an effective perspective in active domains modeling and for the derivation of domain grammars. They show how standard ORM can be extended to an object-role calculus (ORC), including temporal concepts and constraints that enable the modeling of active domains. A suggestion for graphical representation is also provided. The authors hope to contribute to the integration of domain models and viewpoints in an academic and educational context rather than proposing ORM and ORC as new modeling tools in an industrial setting.

Introduction and Background

Conceptual modeling methods such as ER (Chen, 1976), EER (Elmasri & Navathe, 1994; Gogolla, 1994), KISS (Kristen, 1994), NIAM (Nijssen & Halpin, 1989), OOSA (Embley, Kurtz, & Woodfield, 1992), and object-role modeling (ORM; Halpin, 2001) have traditionally been developed with the aim of providing conceptual models of database structures. More recently, however, such modeling methods have shown their use for more generic modeling (of the ontology) of domains, leading to models capturing the concepts of a domain in general, as well as an associated language to express rules (such as business rules) governing the behavior of the domain (Proper, Bleeker, & Hoppenbrouwers, 2004; Proper, Hoppenbrouwers, & Weide, 2005; Spyns, 2005; Spyns, Meersman, & Jarrar, 2002).

The above mentioned modeling methods typically take a natural-language-based perspective to the domain to be modeled. In this perspective, the resulting model is regarded as a domain grammar describing the allowed communication about a domain (the universe of discourse). This way of thinking dates back to the ISO (1987) report *Concepts and Terminology for the Conceptual Schema and the Information Base*, and it is at the base of the modeling methods mentioned. A key advantage of such methods is that having a domain grammar at one's disposal enables validation of the model by domain experts since the model can be validated in terms of statements that are close to the language used by these experts. Also, formal approaches to conceptual modeling imply adherence to a formal language; domain grammars (closely resembling signatures in formal logic) are an excellent basis for further formal modeling.

A basic domain grammar can be extended to cover rules (constraints) governing the structure and behavior of or in the domain. When combined with a reasoning mechanism, such a rule language becomes a domain calculus. In the case of ORM, a domain calculus has been presented in the form of Lisa-D (Hofstede, Proper, & Weide, 1993), a formalization of RIDL (Meersman, 1982). In Proper (1994a) and Bloesch and Halpin (1996), a more practical version was introduced (that is, from an implementation point of view) called ConQuer. What Lisa-D and ConQuer have in common is that they exploit the natural character of the domain grammar in the construction of rules (Hofstede, Proper, & Weide, 1997). As a result, the formulation of rules, as well as chains of reasoning expressed by means of these rules, closely resembles natural language. As mentioned, this supports validation of the models produced.

In the use of domain modeling methods, we observe three important trends that fuel our ongoing research activities. First, more and more organizations strive for more mature levels of system development (Paulk, Curtis, Chrissis, & Weber, 1993). One of the steps toward maturity involves better defining of development processes in order to make them more repeatable. This also applies to modeling processes. Some organizations now indeed strive to make modeling processes more explicitly defined with the aim of achieving more repeatable results.

Modeling methods such as ORM (Halpin, 2001), NIAM (Nijssen & Halpin, 1989), OOSA (Embley et al., 1992), KISS (Kristen, 1994), and DEMO (Reijswoud & Dietz, 1999) not only feature a way of modeling, but also have a fairly well-defined and explicit way of working based on natural-language analysis. The way of working (Wijers & Heijes, 1990) of a method is concerned with processes, guidelines, heuristics, and so forth, which are to be used in the creation of models, as opposed to its way of modeling, which refers to the syntax and semantics of the language in which the models are to be expressed. A well-defined and explicit way of working helps achieve a defined and more repeatable modeling process. Even though the ORM conceptual schema design procedure already provides clear guidelines for creating domain models in a repeatable fashion, more work in terms of sound theoretical underpinning and automated support of the (detailed steps of the) modeling process is still called for (Hoppenbrouwers, Proper, & Weide, 2005b) and is one of the main goals underlying our ongoing research. This is not our main focus here, but it is a partial explanation for our preference for ORM.

The second trend fueling our research is the use of controlled languages as the basis for unambiguous communication concerning models and specifications (European Association of Aerospace Industries [AECMA], 2001; Farrington, 1996; Fuchs & Schwitter, 1996; Hoppenbrouwers, 2003; Schwitter, 2004). The essential idea behind a controlled language is to define a subset of a natural language that is rich enough to have a natural, intuitive feel to it, but still restrictive enough so as to avoid ambiguities. The use of a controlled language requires a realistic and nuanced approach. Too often it is simply assumed that any participant in the modeling process will be able to express herself or himself freely and flawlessly in some controlled language. This is hardly ever a tenable assumption. A possible way of working around this is to use the controlled language receptively for regular participants (not productively) and have a capable intermediary (a person, an automated system, or a combination thereof) rephrase the natural-language statements made by the regular participant(s) in the controlled language. Next, the participant(s) who phrased the original (natural language) statements should confirm the validity of the controlled-language statements (Hoppenbrouwers et al., 2005).

We claim that in view of the goals of formal and unambiguous expression, a domain grammar and associated domain calculus provide a good starting point for engineering controlled languages for use in domain modeling. To some extent, a domain calculus already provides a controlled language, though not necessarily a natural one. Stepwise naturalization (using a number of levels of increasingly natural representations) is the approach we follow here; admittedly, we have not achieved fully acceptable naturalness as of yet, but when we do, it will still be based on fully formal underlying structures, thus linking focused and controlled natural representation of concepts with well-constructed formalism. Calculus-like controlled languages can be used to represent domain-specific reasoning steps, providing an additional form of domain knowledge. In Hoppenbrouwers, Proper, and Weide (2005a), an initial study into the use of a domain calculus for such purposes has been reported.

The third trend we observe is the growing need for integrated models underlying a plethora of viewpoints, fueled by the demands of model-driven architecture or MDA (Frankel, 2003; OMG, 2003) and enterprise architecture (Lankhorst, 2005). The unified modeling language (UML; OMG), as well as approaches for enterprise architecture, feature a wide variety of viewpoint-specific diagramming techniques (viewpoints). A generic domain model can provide a common underpinning of these viewpoints, offering a unified domain ontology. An elaboration of this role of domain models has been presented in Proper et al. (2005) and Proper and Weide (2005). However, since we adopt ORM as a generic technique for creating a unified domain ontology, work still needs to be done to bring together concepts for activity modeling and ORM. This effort is the main focus of this chapter. However, let us first elaborate some more on the reason why we use ORM for our purposes.

Importantly, it is not our intention to propose our ORM extension as a new technique for modeling active domains that is to be on par with, for example, UML activity diagrams. We applaud the use of such diagrams as popular viewpoints. Integration at a deeper level is our primary drive here, mostly for academic and educational purposes. Application in industry should eventually be possible but is not one of our objectives at this point. For clarity's sake, the arguments for using ORM in our current investigation are summarized here:

- There is a growing need for integrated models underlying a plethora of viewpoints. Because of its generic conceptual nature, ORM seems a good candidate for use as a generic technique and method for integrating the many more specialized modeling techniques around.

- ORM provides a good existing basis for investigating, defining, and improving ways of working, in combination with formal rigor and soundness with respect to the way of modeling. Getting a better grip on the way of working serves the higher goal of making modeling processes more mature.

- As part of the way of working, a strongly natural-language-oriented approach to elicitation and validation of models is part and parcel of the ORM-ORC (object-role calculus) approach (Hofstede et al., 1997; Hoppenbrouwers et al., 2005b). The logbook principle and the use of controlled language are elements of this approach.

In this chapter, then, we show how the ORM-ORC approach (with the advantages it does, in our view, possess) could be applied to temporal modeling, integrating it in the existing approach. For this purpose, we need the following ingredients:

- A basic yet coherent set of temporal concepts to work with; this could perhaps have been another set than the one used in this chapter. Please

note that fundamental matters concerning the ideal set of temporal modeling concepts are not within the focus of this chapter. For an elaboration
on such issues, see, for example, Allen and March (2003), March and
Allen (2003), and Khatri, Ram, and Snodgrass (2004). We make do with
a small set of concepts fit for our current purposes based on concepts
from established work-flow-related work (Aalst & Hofstede, 2005).

- Given our goal to apply and maintain ORM-style formal rigor in our
  way of modeling, we show that our ORM extension fits our existing
  ORM-ORC formalization. In order to do this, we need to explain some
  key aspects of that formalization. This does require both active aspects
  (activities, tasks, processes, etc.) and static aspects (results, documents,
  actors, tangible objects, etc.) to be expressed as objects playing roles in
  the domain.

- Given the elicitation and validation goals underlying the ORM-ORC
  approach, we include both a verbal and a graphical style of representation. While both are demonstrated, they are still somewhat experimental.
  They serve to illustrate the integration aspect of our exercise rather than
  as a proposition for new forms of representing active domain models in
  an industrial context.

The body of this chapter is structured as follows:

1. We present an explanation of how, by means of the logbook paradigm,
    the activities taking place in an active domain can be reported in terms
    of (elementary) facts, which can consequently be used (in principle by
    employing ORM's standard approach) to derive a domain grammar.

2. We continue by explaining how any constraints, temporal dependen-
    cies, and so forth governing the flow of activities in a domain can then
    be formulated using a domain calculus referred to as the object-role
    calculus.

3. Special graphical conventions are introduced to provide more compact
    representations of specific aspects of the active domain, such as the flow
    of activities, or the involvement of actors.

4. We conclude the chapter.

The Logbook Paradigm

When focusing on active domains, ORM needs to be refined in order to better cater to the active aspects of such domains. The underlying challenge is to extend ORM to be able to deal with such domains while at the same time maintaining ORM's natural-language-based modeling rigor. In doing so, we base ourselves on earlier (partial) results (Frederiks, 1997; Frederiks & Weide, 2002; Proper, 1994b; Proper & Weide, 1994).

Modeling an active domain requires a modeling language to deal with the notion of time. In the past, ORM had already been extended with the concept of time and evolution (Proper, 1994b; Proper & Weide, 1994). In this chapter we propose a formalization of temporal concepts in terms of a logbook (Frederiks, 1997; Frederiks & Weide, 2002), which is intended to trace or mirror the activities taking place in the domain. Such a logbook will consist of a series of events reporting on the life cycle of facts in the domain. The following is an example.

Traffic light 20 is green
ceased being true at 11:03:20 on 22-05-2006
Employee John works on the completion of order 50
started being true at 09:30 on 19-05-2006

The logbook approach is a natural extension of the earlier discussed natural-language-based perspective on modeling. To be more precise, we regard a history as the collection of events that have taken place in the domain, while a logbook is a description of such a history. The facts contained in the descriptions of the events are assumed to be expressed in terms of some seminatural-language (controlled language) sentences as is normally the case in ORM's way of working. Using a traditional ORM approach, the set of facts used and allowed in a logbook can be generalized to a set of fact types, which together comprise the ORM model underlying the domain. This model then defines the domain grammar of the controlled language in which the facts are to be formulated.

Traditionally, ORM is based on the modeling of facts in general. In the context of an active domain, these facts correspond to statements about what is the case and/or has happened in the domain at specific points in time. In ORM, the actual modeling process starts out from the verbalization of such facts. These verbalizations are the starting point for the creation of the domain grammar. When considering an active domain, the set of facts that can be reported about this domain fall into two categories: (a) acts reporting on the performance of actions and (b) effects reporting the results of actions (note that these two classes correspond to what Dietz, 2005, respectively refers to as *acta* and *facta*). This dichotomy applies at the instances level (the facts) as well as the type level (the fact types), leading to act types and effect types respectively as subclasses of fact types. In the case of acts, the objects involved (i.e., playing a role in the act) can be classified further into actors (objects responsible for performing the act) and actands (objects that are the effect of the act).

We assume that each event described in the logbook and the objects participating in the event can be uniquely identified in that logbook. We will call this the event identification principle. It does not inhibit different events to occur at the same time. In order to distinguish between accidentally and necessarily coupled events, we assume that events may also have a compound nature in such a way that (a) different events in a logbook are independent of each other, and (2) events cannot be split into multiple independent events.

We take the perspective that the state of an active domain is the result of the sequence of actions leading up to that state. These actions may either take place in the domain, or outside the domain (the latter possibility includes the very creation of the domain). As a result, we take the position that the effects can in principle be derivable from the set of reported acts. This is what we call the action dominance principle. This principle leads to the theoretical question of how persistent properties, such as the speed of light, are to be treated in our logbook approach. This is covered by the property origination principle, which states that each domain property pertains to (a) some act that took place in the domain, (b) some effect of some act in the domain, or (c) some effect of the domain's creation (i.e., the result of a "big bang" act). As a consequence, at each moment the state of the system is (in principle) the result of all the effects of the domain's creation and the acts that were reported since then.

An important consequence of the property origination principle is that (for most objects in the domain) the property of being alive should be the result of some act (this notion is similar to that of "existence time" as discussed in Khatri et al., 2004, among others). Therefore, objects that are not present in the initial state in principle require an explicit birth event. This is called the birth principle. Obviously, an object cannot be responsible for its own birth as it cannot be active before coming into existence. The consequence is that some other object has to be responsible for causing the event, thus playing a dominant role in it. If the existence of an object may terminate, then there should in principle be an explicit death action that enforces an object to have the property of being dead.

An immediate consequence of the birth principle and the event identification principle is that objects may be identified by their birth event. If an event starts life for more objects, then we require that the individual objects in this case may be identified by this event in addition to the role they play in the event.

Note that the above principles could only hold absolutely under a closed-world assumption, which in most practical cases is naïve. We therefore emphasize that more traditional means of object identification are not excluded from our approach. The principles as presented merely reflect our perspective on active domains. For example, if the birth event of an object is unknown, or even if it is known, identification of the object as such, by means of a simple key or label, would still be quite acceptable. Analytically, however, questions may be raised as to the origin or history of the object. Whether such questions are acutely relevant depends on the modeling context.

Object-Role Calculus

This section concerns a conceptual language in which rules can be expressed for describing the behavior that may be observed in a logbook compatible with the domain being modeled. The language presented, referred to as ORC, is a variant of Lisa-D (Hofstede et al., 1993), a formalization of RIDL (Meersman, 1982). Lisa-D was originally designed to describe all computable sets of facts that can be derived from the elementary facts defined in some underlying conceptual schema. The conceptual schema specifies all elementary sentences applicable in a domain. The semantics of Lisa-D have been described in terms of multisets. In this chapter, we will provide a light-weight definition of the ORC variant of Lisa-D, which is intended to describe temporal and static aspects of the underlying domain.

Grounding in Temporal Logic

The semantics of ORC are grounded on Kripke structures (Chellas, 1980). A Kripke structure basically is a graph whose nodes represent the reachable states of some system and whose edges represent state transitions. A labeling function maps each node to a set of properties that hold in the corresponding state. Temporal logics are traditionally interpreted in terms of Kripke structures.

An application domain, then, is seen as a Kripke structure $\langle S.R.s_0, \prod, L\rangle$, where:

1. S is a nonempty set of states,
2. $R \subseteq S \times S$ is a total transition function, that is, $\forall_s \exists_t [(s,t) \, R]$,
3. s_0 is the initial state,
4. $\prod$ is a nonempty set of atomic propositions, and
5. L is a labeling function that maps each state on a subset of $\prod$.

Our main assumption is that the state of an application domain is in principle described by its history so far. As a consequence, a state corresponds uniquely to a logbook. Hence, the transition function extends a logbook with a new event description, and the initial state corresponds to the empty logbook.

From the structure of the events in the logbook, the elementary object types can be derived. Their possible instantiations form the set of $\prod$ atomic propositions. The labeling function L then assigns the population of object types that is constructed by a logbook.

A linear-time temporal logic is syntactically described by the following BNF grammar (see, for example, Lipeck & Saake, 1987):

$$\varphi \rightarrow true \mid false \mid \prod \mid \neg\varphi \mid q \wedge \varphi \mid \varphi \vee \varphi \mid \varphi \Rightarrow \varphi \mid \mathsf{X}\varphi \mid \mathsf{F}\varphi \mid \mathsf{G}\varphi \mid \varphi \, \mathsf{U}\psi.$$

The expression $\mathsf{X}\varphi$ states that φ will hold in the next state, $\mathsf{F}\varphi$ means that φ will eventually hold, $\mathsf{G}\varphi$ means that φ will globally hold, and $\varphi \, \mathsf{U}\psi$ states that at some point ψ will hold while in all states before, φ is valid. Let M be a Kripke structure over logbook LB, and let σ be a history. We will further assume an environment E for evaluation, consisting of a partial assignment of values to a set V of variables. The standard semantic interpretation of the temporal operators is as follows; for lack of a typographic alternative, we use the "$\equiv$" symbol here for "is defined as."

$$M, E, \sigma \models \mathsf{X}\varphi \qquad \equiv \qquad M, E, \sigma^i \models \varphi$$

$$M, E, \sigma \models \varphi\, \mathsf{U}\psi \qquad \equiv \qquad \exists_n [\, \forall_{0<i<n} [\, M, E, \sigma^i \models \varphi\,] \wedge M, E, \sigma^n \models \psi\,],$$

where σ^i denotes the ith element of sequence σ, and σ^i the subsequence of σ starting at position i. The other temporal operators are defined in terms of these base operators: $\mathsf{F}\varphi$ is equivalent with *true* $\varphi\,\mathsf{U}\psi$ and $\mathsf{G}\varphi$ is defined as $\neg\,\mathsf{F}\,\neg\,\varphi$. The propositional operators are also interpreted in the standard way:

$$M, E, \sigma \models \neg\varphi \qquad \equiv \qquad \neg\, M, E, \sigma \models \varphi$$

$$M, E, \sigma \models q \wedge \psi \qquad \equiv \qquad M, E, \sigma \models \varphi \wedge M, E, \sigma \models \psi.$$

The constant *false* is introduced as $p \wedge \neg p$, where p is any proposition from $\prod$, and *true* is derived by $\neg false$. The other logical operators ($\vee$ and $\Rightarrow$) are defined in the usual way. The conversion from a temporal proposition to a static expression requires the evaluation of the static expression for the population $L(\sigma(0))$ at the required point in time. This will be further elaborated later.

Historical Information Descriptors

History descriptors in ORC are meant to provide a language construct for reasoning about the application domain in a historical setting. For the purpose of this chapter, it will be sufficient to make direct transcriptions of the basic temporal operators. For this, the syntactical construct of history descriptor is introduced. Let H be a history descriptor, then the semantics of H are denoted as $[(H)]$:

$$[(\text{always } H)] \qquad \equiv \qquad \mathsf{G}\,[(H)]$$

$$[(\mathsf{X}\,H)] \qquad \equiv \qquad \mathsf{X}\,[(H)].$$

In addition, we introduce the following abbreviations:

sometimes H	$\equiv$	$\neg$ always $\neg H$
H_1 precedes H_2	$\equiv$	always$((\mathsf{F}H_1)\ \mathsf{U}\ H_1)$
H_1 during H_2	$\equiv$	always$(H_1 \Rightarrow H_2)$
H_1 triggers H_2	$\equiv$	always$(H_1 \wedge \neg H_2 \Rightarrow \mathsf{X}\,(\neg H_1 \wedge \neg H_2))$

We will now introduce some example rules that match a more elaborate example concerning an educational organization. The first rule will be a target for the educational organization. The second rule describes a trigger that, whenever the condition $H_1 \wedge \neg H_2$ is met, will respond by setting the condition $\neg H_1 \wedge \neg H_2$ at the next moment.

sometimes Lecturer lectures Course

Lecturer sets up Course precedes Lecturer lectures Course

This latter expression, however, is misleading as it does not bring about a connection between some specific lecturer and some specific course being set up and being lectured. In natural language, demonstratives (for example, *this* or *that* in English) are used in most cases to make such references. We therefore introduce the following:

$x\ [[\ D_1\ \text{PRECEDES}\ D_2\ ]]\ y$	$\equiv$	$(x\ [[D_1]]\ y)$ precedes $\exists_z\ [\ z\ [[D_2]]\ y]$
$x\ [[\ D_1\ \text{DURING}\ D_2\ ]]\ y \equiv$		$(x\ [[D_1]]\ y)$ during $\exists_z\ [\ z\ [[D_2]]\ y]$

The semantics and syntax of these constructions are further explained later. Please note that we use the repeated bracket "[[" notation here for typographical lack of a properly fused double square bracket. All immediately adjoining brackets in this chapter are double square brackets, never two single square brackets.

Demonstrative Descriptors

The main idea behind ORC, as present in its early ancestor RIDL (Meersman, 1982), is a functional, variableless description of domain-specific properties (and queries). RIDL does contain a linguistic reference mechanism (the demonstrative THAT). In ORC, variables have been introduced to handle more subtle referential relations that cannot be handled by demonstratives. Variables are special names that are instantiated once they are evaluated in a context that generates values for these variables. The concept of environment is used to administrate the value of variables. In environment E, the variable v will evaluate to $E(v)$. Some examples of the use of variables follow:

Lecturer:x being hired precedes x sets up Course

Lecturer:x sets up c precedes x lectures Course:c

In this example, the expression Lecturer:x is a defining occurrence of variable x in which Lecturer has the role of value generator. The environment is used to administrate the variable-value assignment (see Hofstede et al., 1993, for more details).

Information Descriptors

The syntactic category used to retrieve a collection of facts is called the information descriptor. We will discuss the semantics of elementary information descriptors and briefly summarize the construction of information descriptor (a diagram is provided in Figure 1; for more details, see Hofstede et al., 1993). Information descriptors are constructed from the names of object types and role types. The base construction for sentences is juxtaposition. By simply concatenating information descriptors, new information descriptors are constructed.

Information descriptors are interpreted as binary relationships; they provide a binary relation between instances of the population induced from the history. The semantics of information descriptor D is denoted as $[[D]]$; we will write $x\ [[D]]\ y$ to denote the relationship between x and y. The statement M, E, $\sigma \models x\ [[D]]\ y$ asserts that for Kripke structure M in environment E from history σ, the relationship $x\ [[D]]\ y$ can be derived..

Figure 1. Role names

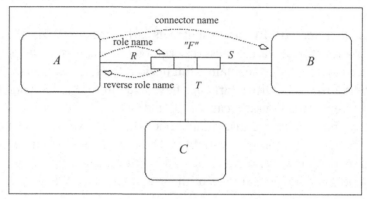

A population assigns to each object type its set of instances. Let n be the name of object type N, and r the name of a role type R; then n and r are information descriptors with the following semantics:

$$M, E, \sigma \models x \ [[n]] \ y \ \equiv \ x \in L(\sigma (N)) \wedge x = y$$
$$M, E, \sigma \models x \ [[r]] \ y \ \equiv \ (x,y) \in L(\sigma (R)).$$

A single role may, in addition to its "normal" name, also receive a reverse role name. Let v be the reverse role name of role R; then we have:

$$M, E, \sigma \models x \ [[v]] \ y \ \equiv \ (y, x) \in L(\sigma (R)).$$

A combination of roles involved with a fact type may receive a connector name. The connector name allows us to "traverse" a fact type from one of the participating object types to another one. If c is the connector name for a role pair $\langle R, S \rangle$, then the semantics of the information descriptor c is defined as:

$$M, E, \sigma \models x \ [[c]] \ z \ \equiv \ \exists_y [M, E, \sigma \models x \ [[R]] \ y \wedge M, E, \sigma \models y \ [[S]] \ z].$$

Elementary information descriptors can be composed into complex information descriptors using constructions such as concatenation, conjunction, implication, disjunction, and complementation. These may either refer to the fronts alone or to both fronts and tails of descriptors. For more details, see Hofstede et al. (1993). In this chapter we use:

$$x \, [[D_1 \, D_2]] \, y \qquad \equiv \qquad \exists_z \, [\, x \, [[D_1]] \, z \wedge z \, [[D_2]] \, y \,]$$

$$x \, [[D_1 \text{ AND ALSO } D_2]] \, y \qquad \equiv \qquad \exists_z \, [\, x \, [[D_1]] \, z \,] \wedge \exists_z \, [x \, [[D_2]] \, z \,] \wedge x = y,$$

where D_1 and D_2 are information descriptors, and x, y, and z are variables. Some example expressions would be the following:

Person working for Department "I&KS"

Persons working for department "I&KS"

Person (working for Department "I&KS" AND ALSO owning Car of Brand "Seat")

Persons working for department "I&KS" who also own a car of brand Seat

Note that the natural-language likeness of the ORC expressions used in this chapter can yet be improved considerably. In the above example, we have added a naturalized version of the ORC expression in italics. We are in the process of developing a formal grammar for a naturalized version of ORC that has a 1:1 correspondence to basic (deep) ORC structures. However, because this grammar is not available as of yet, we provide ad hoc naturalized expressions for clarification.

Rules

ORC has a special way of using information descriptors to describe rules that should apply in a domain (note that constraints are in fact rules). Rules consist of information descriptors that are interpreted in a Boolean way; that is, if no tuple satisfies the relationship, the result is false, and otherwise it is true. Some examples of such constructions are as follow:

$$[[\text{SOME } D]] \quad \equiv \quad \exists_{x,y} [\, x \, [[D]] \, y \,]$$
$$[[\text{NOT } R_1]] \quad \equiv \quad \neg[[R_1]]$$
$$[[\text{NO } D]] \quad \equiv \quad [[\text{NOT SOME } D]],$$

where D is an information descriptor and R_1 a rule.

Graphical Representation

Currently, we are experimenting with the effective graphical representation of some key classes of temporal dependencies. In Proper et al. (2005), we have provided some examples using notations inspired by the field of work-flow modeling (Aalst & Hofstede, 2005).

A key modeling construct is the notion of a life-cycle type. An example of its use is provided in Figure 2, which contains two interlinked life-cycle types: Course Offering and Course Attendance. Each of these life-cycle types comprises multiple action types.

Figure 2. Lecturing example

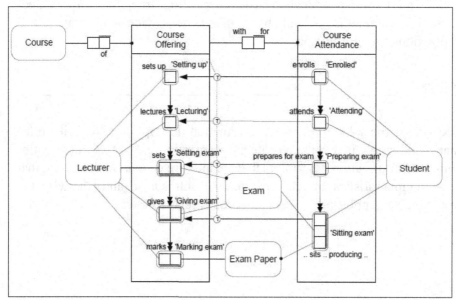

In the example domain, courses are offered to students. In offering a course, a lecturer starts by setting up the course offering. This is followed by the actual lecturing. After lecturing the course, the lecturer sets an exam. This exam is given to the students attending the course, after which the lecturer marks the exam papers produced by the students. Students attend the course by enrolling. After their enrollment, they attend the course. Once the course is finished, they prepare themselves for the exam, which is followed by the actual exam, leading to an exam paper.

In general, the life-cycle type typically involves multiple action types and can best be regarded as an abbreviation as illustrated in Figure 3. The temporal dependency between x and y is defined as follows:

$$x \longrightarrow\!\!\!\!> {}_S y \quad \equiv \quad x \text{ being act of S PRECEDES } y \text{ being act of } S.$$

The enrollment by students in a course should take place during the setup phase of a course. This is enforced by means of the temporal subset constraint from the Enrolling action type to the Setting Up action type. The connection between the temporal subset constraint and the Course Offering life-cycle type signifies that the temporal subset constraint should be evaluated via this object type. In general, the semantics are expressed as:

$$x \subseteq_T y \quad \equiv \quad x \text{ DURING } y.$$

Figure 3. Life-cycle types

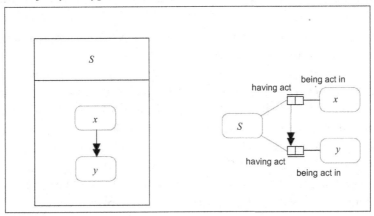

Figure 4. Lecture activities

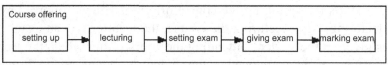

In the case of Figure 2, we have specified a join path, leading, for example, to the following:

Enrolling being act of Course Attendance for Course Offering
DURING

Setting up being act of Course Offering

Enrolling (which is an act of course attendance, in response to course offering)

takes place during

setting up (which is an act of course offering)

Finally, a model as presented in Figure 2 can be used as a basis for deriving specialized views such as depicted in Figure 4, focusing on the flow of activities performed by a lecturer.

Conclusion

The research reported in this chapter is part of our effort to find a suitable generalized domain modeling method to model active domains in view of an ongoing attempt to achieve integrated domain ontologies underlying the many viewpoints in conceptual modeling. In this chapter, we have proposed the application of ORM rigor and the use of the ORM approach to model elicitation and validation in modeling active domains. We have introduced the logbook paradigm as a history-oriented extension of the traditional natural-language orientation of ORM. To be able to define rules governing the behavior of active domains, we have introduced ORC. The semantics of this rule language has been defined in terms of Kripke structures. Finally, we have shown how ORM can be extended with graphical constructs, in particular life-cycle types, focusing on temporal dependencies in a domain. This notation allows us to also derive specific views on a domain focusing solely on temporal behavior, which has been demonstrated.

As made clear earlier, we do not put forward the verbal and graphical notations presented in this chapter as a competitor to existing and well-established techniques for modeling active domains. It is integration we strive for, and we do view ORM and ORC as good candidates for providing a foundation for the fundamental integration of many existing, dedicated models and views.

Validation of our representations in an industrial context seems not quite relevant, and has not been attempted. However, in academic education, ORM, ORC, and recently the temporal extension presented in this chapter have been successfully used to teach MSc students in information science the fundamentals of formal conceptual modeling. We found it very helpful indeed to present students with an integrated set of models firmly grounded in a well-understood formalism, aiding them in coming to terms with the many complex issues involved (both formal and methodological). In addition, our experience is that once the fundamentals have been acquired, students can easily apply them to other modeling techniques and methods, and learn and understand these better and more quickly than their colleagues did some years previous when an integrated foundation was still lacking in the curriculum (other modeling techniques are in fact still taught). Admittedly, these experiences have so far not been backed up by systematic research. Still, we consider the results good enough to continue our approach and further develop integrated, ORM-style conceptual modeling as a core around which other modeling techniques and viewpoints are positioned.

As a next exercise, we intend to take some typical patterns from, for example, enterprise modeling and work-flow modeling, and study how to ground them in terms of an underlying ORM domain model with accompanying ORC rules. We expect this to provide further progress in our effort to find a suitable generalized domain modeling method to model active domains.

References

Aalst, W. van der, & Hofstede, A. ter. (2005). YAWL: Yet another workflow language. *Information Systems, 30*(4), 245-275.

Allen, G. N., & March, S. T. (2003). Modeling temporal dynamics for business systems. *Journal of Database Management, 14*(3), 21-36.

Bloesch, A., & Halpin, T. (1996). ConQuer: A conceptual query language. In B. Thalheim (Ed.), *Proceedings of the 15th International Conference on Conceptual Modeling (ER'96)* (LNCS 1157, pp. 121-133). Berlin, Germany: Springer.

Chellas, B. (1980). *Modal logic: An introduction.* Cambridge, United Kingdom: Cambridge University Press.

Chen, P. (1976). The entity-relationship model: Towards a unified view of data. *ACM Transactions on Database Systems, 1*(1), 9-36.

Dietz, J. L. (2005). A world ontology specification language. In R. Meersman, Z. Tari, & P. Herrero (Eds.), *On the Move to Meaningful Internet Systems 2005: OTM Workshops. OTM Confederated International Workshops and Posters, AWeSOMe, CAMS, GADA, MIOS+INTEROP, ORM, PhDS, SeBGIS, SWWS, and WOSE 2005* (LNCS 3762, pp. 688-699). Berlin, Germany: Springer-Verlag.

Elmasri, R., & Navathe, S. (1994). Advanced data models and emerging trends. In *Fundamentals of database systems* (2nd ed., chap. 21). Redwood City, CA: Benjamin Cummings.

Embley, D., Kurtz, B., & Woodfield, S. (1992). *Object-oriented systems analysis: A model-driven approach.* New York: Yourdon Press.

European Association of Aerospace Industries (AECMA). (2001). *AECMA simplified English: A guide for the preparation of aircraft maintenance documentation in the international maintenance language* (Issue 1, Revision 2). Retrieved from http://www.aecma.org

Farrington, G. (1996). *An overview of the international aerospace language.*

Frankel, D. (2003). *Model driven architecture: Applying MDA to enterprise computing.* New York: Wiley.

Frederiks, P. (1997). *Object-oriented modeling based on information grammars* [doctoral dissertation]. Nijmegen, the Netherlands: University of Nijmegen.

Frederiks, P., & Weide, T. van der. (2002). Deriving and paraphrasing information grammars using object-oriented analysis models. *Acta Informatica, 38*(7), 437-488.

Fuchs, N., & Schwitter, R. (1996). Attempto controlled English (ACE). *Proceedings of the First International Workshop on Controlled Language Applications (CLAW96)*, 124-136.

Gogolla, M. (1994). *An extended entity-relationship model: Fundamentals and pragmatics* (LNCS 767). Berlin, Germany: Springer.

Halpin, T. (2001). *Information modeling and relational databases: From conceptual analysis to logical design.* San Mateo, CA: Morgan Kaufmann.

Hofstede, A. ter, Proper, H. A., & Weide, T. van der. (1993). Formal definition of a conceptual language for the description and manipulation of information models. *Information Systems, 18*(7), 489-523.

Hofstede, A. ter, Proper, H. A., & Weide, T. van der. (1997). Exploiting fact verbalisation in conceptual information modelling. *Information Systems, 22*(6/7), 349-385.

Hoppenbrouwers, S. (2003). *Freezing language: Conceptualisation processes in ICT supported organizations* [doctoral dissertation]. Nijmegen, the Netherlands: University of Nijmegen.

Hoppenbrouwers, S., Proper, H. A., & Weide, T. van der. (2005a). Fact calculus: Using ORM and Lisa-D to reason about domains. In R. Meersman, Z. Tari, & P. Herrero (Eds.), *On the Move to Meaningful Internet Systems 2005: OTM Workshops. OTM Confederated International Workshops and Posters, AWeSOMe, CAMS, GADA, MIOS+INTEROP, ORM, PhDS, SeBGIS, SWWS, and WOSE 2005* (LNCS 3762, pp. 720-729). Berlin, Germany: Springer-Verlag.

Hoppenbrouwers, S., Proper, H. A., & Weide, T. van der. (2005b). A fundamental view on the process of conceptual modeling. In L. Delcambre, C. Kop, H. Mayr, J. Mylopoulos, & O. Pastor (Eds.), *Conceptual Modeling: ER 2005. 24th International Conference on Conceptual Modeling* (LNCS 3716, pp. 128-143). Berlin, Germany: Springer-Verlag.

ISO. (1987). *Information processing systems: Concepts and terminology for the conceptual schema and the information base* (ISO/TR 9007:1987). Retrieved from http://www.iso.org

Khatri, V., Ram, S., & Snodgrass, R. T. (2004). Augmenting a conceptual model with geospatiotemporal annotations. *IEEE Transactions on Knowledge and Data Engineering, 16*(11), 1324-1338.

Kristen, G. (1994). *Object orientation: The KISS method. From information architecture to information system.* Reading, MA: Addison Wesley.

Lankhorst, M. (Ed.). (2005). *Enterprise architecture at work: Modelling, communication and analysis.* Berlin, Germany: Springer.

Lipeck, U., & Saake, G. (1987). Monitoring dynamic integrity constraints based on temporal logic. *Information Systems, 12*(3), 255-269.

March, S. T., & Allen, G. N. (2003). On the representation of temporal dynamics. *Advanced Topics in Database Research, 2,* 37-53.

Meersman, R. (1982). *The RIDL conceptual language* (Tech. Rep.). Brussels, Belgium: International Centre for Information Analysis Services, Control Data Belgium, Inc.

Nijssen, G., & Halpin, T. (1989). *Conceptual schema and relational database design: A fact oriented approach.* Englewood Cliffs, NJ: Prentice-Hall.

OMG. (2003). *UML 2.0 superstructure specification: Final adopted specification* (Tech. Rep. No. ptc/03-08-02). Retrieved from http://www. omg.org

Paulk, M., Curtis, B., Chrissis, M., & Weber, C. (1993). *Capability maturity model for software* (Version 1.1, Tech. Rep. No. SEI-93-TR-024). Pittsburgh, PA: Software Engineering Institute, Carnegie Mellon University.

Proper, H. A. (1994a). *ConQuer-92: The revised report on the conceptual query language LISA-D* (Tech. Rep.). Brisbane, Queensland, Australia: Asymetrix Research Laboratory, University of Queensland.

Proper, H. A. (1994b). *A theory for conceptual modelling of evolving application domains* [doctoral dissertation]. Nijmegen, the Netherlands: University of Nijmegen.

Proper, H. A., Bleeker, A., & Hoppenbrouwers, S. (2004). Object-role modelling as a domain modelling approach. In *Proceedings of the Workshop on Evaluating Modeling Methods for Systems Analysis and Design (EMMSAD '04), held in conjunction with the 16th Conference on Advanced Information Systems 2004 (CAiSE 2004)* (Vol. 3, pp. 317-328).

Proper, H. A., Hoppenbrouwers, S., & Weide, T. van der. (2005). A fact-oriented approach to activity modeling. In R. Meersman, Z. Tari, & P. Herrero (Eds.), *On the Move to Meaningful Internet Systems 2005: OTM Workshops. OTM Confederated International Workshops and Posters, AWeSOMe, CAMS, GADA, MIOS+INTEROP, ORM, PhDS, SeBGIS, SWWS, and WOSE 2005* (LNCS 3762, pp. 666-675). Berlin, Germany: Springer-Verlag.

Proper, H. A., & Weide, T. van der. (1994). EVORM: A conceptual modelling technique for evolving application domains. *Data & Knowledge Engineering, 12*, 313-359.

Proper, H. A., & Weide, T. van der. (2005). Modelling as selection of interpretation. In *Modellierung 2006* (LNI P82, pp. 223-232).

Reijswoud, V. van, & Dietz, J. (1999). *DEMO modelling handbook* (2nd ed., Vol. 1). Delft, the Netherlands: Delft University of Technology.

Schwitter, R. (2004). *Controlled natural languages*. Centre for Language Technology, Macquary University. Retrieved from http://www.ics.mq.edu.au/rolfs/controlled- natural- languages/

Spyns, P. (2005). Object role modelling for ontology engineering in the DOGMA framework. In R. Meersman, Z. Tari, & P. Herrero (Eds.), *On the Move to Meaningful Internet Systems 2005: OTM Workshops. OTM Confederated International Workshops and Posters, AWeSOMe, CAMS, GADA, MIOS+INTEROP, ORM, PhDS, SeBGIS, SWWS, and WOSE 2005* (LNCS 3762, pp. 710-719). Berlin, Germany: Springer-Verlag.

Spyns, P., Meersman, R., & Jarrar, M. (2002). Data modelling versus ontology engineering. *ACM SIGMOD Record, 31*(4), 12-17.

Wijers, G., & Heijes, H. (1990). Automated support of the modelling process: A view based on experiments with expert information engineers. In B. Steinholz, A. Sølvberg, & L. Bergman (Eds.), *Proceedings of the Second Nordic Conference CAiSE '90 on Advanced Information Systems Engineering* (LNCS 436, pp. 88-108). Berlin, Germany: Springer.

Chapter VI

Method Chunks to Federate Development Processes

Isabelle Mirbel, I3S Laboratory, France

Abstract

Method engineering aims at providing effective solutions to build, improve, and support the evolution of development methodologies. Contributions in the field of situational method engineering aim at providing techniques and tools allowing one to build project-specific methodologies. However, little research has focused on how to tailor such situational methodologies when used as organization-wide standard approaches. Moreover, current approaches have been thought of for method engineers, that is to say, expert users, and they are not enough dedicated to nonexpert ones. In this context, we propose an approach that consists of federating the method chunks built from the different project-specific methods in order to allow each project to share its best practices with the other projects without imposing to all of them a new and unique organization-wide method.

Introduction

Several decades of work have been spent to provide effective solutions to build, improve, and support the evolution of development methodologies. Different approaches have been successively proposed to provide suitable support to software-based information system development. Experiments show that the provided models and methodologies have been adapted to each of the different situations in which they have been used. At the end, almost every project has carried out tailoring in order to apply effectively best standard practices. There exist now a lot of variations around a given methodology, each of them appearing suitable for the situation (i.e., the organization or the project) it has been customized for, but they are not so easily translatable in a somewhat different situation, even inside the same domain (i.e., the application domain or the organization).

A development methodology (or process) may be seen as a transformation process (where nonformal specifications are transformed into more formal specification and then code), a decision-making process (where the taken decisions are recorded all along the development process), or a problem-solving process (where solutions are provided to the successive problems encountered during the development process). Especially with regard to these two last viewpoints (decision-making and problem-solving aspects), it would be interesting to benefit from the experiences acquired during the resolution of previous problems. Moreover, the rich and long experience we already have in supporting software-based information system development leads us to try to capitalize and share best practices in this field as it has already been successfully done in the software development domain.

Indeed, there is a need for the capitalization and sharing of knowledge about method engineering as well as a need for customization and tailoring of this knowledge to be better adapted to the organization, the project it is deployed in, and even the user it is targeted for. In this chapter, we start first by discussing the proposals made in the field of method engineering (and especially situational method engineering, aiming at providing solutions to customize development methodologies) and the work done on software reuse. Then we show the shortcomings of the provided approaches. As will be detailed, current approaches have been thought of for expert users and not enough are dedicated to nonexpert ones. Moreover, they are not very suitable when used as organization-wide standards.

In order to overcome the identified shortcomings of current approaches and to provide, in the same framework, means to support the customization of development methodologies at different levels (organization, project, users), and means to capitalize and share knowledge about method engineering, one solution is to capture and understand all the situational methodologies used inside each project of the organization to build an organization-wide standard method by merging the best practices coming from each of them. This solution requires that method engineers, that is to say, the persons in charge of building and deploying the development methodology in the organization, are able to capture and understand each variation of each methodology in each project, which is not an easy task. It also requires that each method user (i.e., the project member applying or using the methodology) accept and use the new organization-wide method instead of his or her customized version of it, which is also not easy. In this context, we propose an alternative solution that consists of federating the different development methodologies used inside each project (that we call project-specific development methodologies) in order to allow each project to share its best practices with the other projects without imposing to all of them a unique organization-wide development methodology.

Our proposal is presented later. First, we start by discussing the two main elements of our environment:

- A method-chunk repository consisting of a number of adequately described fragments that have been identified and extracted from existing development methodologies
- A reuse frame consisting of a number of keywords meaningful and discriminant to describe method fragments. These keywords are organized into a tree structure that allows structured storage and subsequent retrieval of fragments.

Then, we discuss how to support development process federation thanks to the method-chunk repository and the reuse-frame structure. We first explain how to qualify method fragments with the help of the keywords provided in the reuse frame and how to clarify method-user needs in terms of methodological support, again with the help of the reuse-frame content. Then we present the similarity metrics and the closeness distance we propose to quantify the similarity between method fragments and between method fragments and user needs. Finally, we discuss the perspective of our work and conclude.

Background

The software-based information system development field has always been very demanding in techniques and methodologies to enhance the quality of its products and the performance of its processes. It has led to the proposal of numerous models, methodologies, and associated tools. It has also resulted in a rich know-how that conducted our community to consider software-based information system development from the reuse perspective. As our interest is in method engineering and in software-based information systems, in the following, we discuss proposals that emanate from these two research domains.

In the field of method engineering, whose aim is to provide effective solutions to build, improve, and support the evolution of software-based information system development methodologies, proposals have been made to reuse know-how about software-based information system development methodologies in order to build new development methodologies better adapted to the features of a specific project. These approaches are reassembled under the term situational method engineering. They are presented later.

Work in the field of software reuse also includes attempts to apply reuse to the products and processes development engineering is made of. Some proposals also deal with method-engineering knowledge reuse.

Situational Method Engineering

Different approaches have been successively proposed to provide suitable support to software-based information system development. Software-based information system development methodologies (abbreviated to development methodologies in the following) have been formalized to support different intentions (Leppänen, 2006). Some of them aim at formalizing the development process as a transformation process; others lead to capture the development process as a decision-making process; there are also approaches formalizing the development process as a problem-solving process. Recent approaches try to apprehend it as a learning process.

Indeed, whatever the followed intention is, development methodologies have been developed with a broad scope of situations in mind and finally seem too generic to be applied as such in a specific project. Projects differ with respect to their development context, delivery, project team, deadline, and so forth.

Even similar projects require different levels of tailoring due to differences in the organizational structure. Also, an organization may consist of projects having significantly different characteristics, and therefore requiring different development methodologies. Almost every organization or project carries out tailoring in order to apply effectively best standard practices. There is a need to adapt practices to suit the varying needs of projects and organizations. Several studies have proposed factors influencing process tailoring including domain characteristics, project characteristics, project goals and assumptions, organizational structure, corporate size, maturity level, and so forth (Ginsberg & Quinn, 1995; Mirbel & de Rivières, 2002; van Slooten & Hodes, 1996).

Situational method engineering aims at building specific development methodologies to meet the requirement of a particular project situation by reusing and assembling parts of existing methodologies (Brinkkemper, Saeki, & Harmsen, 1998; Harmsen, 1997; Ralyté, 2001). In these kinds of approaches, the method engineer is responsible for building the fragment repository after having identified and extracted the reusable parts of existing methodologies or having generated them from a metamodel. Then, the method engineer is in charge of building a new and adapted methodology by assembling the suitable reusable parts stored in the fragment repository.

Based on the observation that any method has two interrelated aspects, product and process, several authors propose two types of method fragments: process fragments and product fragments (Brinkkemper et al., 1998; Harmsen, Brinkkemper, & Han Oei, 1994; Punter & Lemmen, 1996). Other authors consider only process aspects and provide process components (Firesmith & Henderson-Sellers, 2002) or process fragments (Mirbel, 2004). Others integrate these two aspects in the same module, called a method chunk (Ralyté & Rolland, 2001; Rolland, Plihon, & Ralyté, 1998). The notion of method bloc proposed in Prakash (1999) is similar to the method chunk as it also combines product and process perspectives into the same modeling component. An agent-oriented approach combining product and process perspectives is also proposed in Cossentino and Seidita (2004). Another kind of situational method engineering approach is based on generic conceptual patterns for method construction and extension instead of fragments (Rolland & Plihon, 1996). Conceptual patterns capture generic laws governing the construction of different but similar development methodologies. Deneckère and Souveyet (1998) propose a domain-specific process and product patterns for existing method extension. Decision-making patterns capturing the best practices in enterprise modeling are proposed in Rolland, Nurcan, and Grosz (2000) to better support development processes.

Indeed, different objectives are targeted by fragments (or components or chunks) in the method engineering literature. A first family of approaches aims at documenting development methodologies through well-defined fragments (Storrle, 2001). These approaches do not provide powerful supports, nor do they reuse the fragments from one development methodology to another, nor do they customize the development methodology for a specific project or organization. Their strength resides in the effort of specification with regard to the elements a development methodology is made of (tasks, activities, resources, etc.). A second family of approaches groups works whose aim is to help in building new development methodologies starting from existing ones (instead of building them from scratch; Brinkkemper et al., 1998; Ralyté, 2001). In this kind of approaches, the focus is on the operators provided to allow a new combination of existing fragments, and on mechanisms to evaluate the similitude among fragments. Both families of approaches are dedicated to method engineers.

Proposals about situational method engineering mainly provide solutions to allow method engineers to customize a development methodology with regard to the specificities of a particular project in a specific organization. For this purpose, means have been provided to formalize development methodologies and to specify reusable pieces of development methodology that are stored in a dedicated repository. Dedicated operators and processes have also been discussed to support method fragments assembly into new and more suitable development methodologies.

Reuse and Method Engineering

Providing support to development methodologies and especially in the field of situational method engineering means to provide means to capitalize and share best practices in software-based information system development, that is to say, method engineering knowledge. It is recognized as important to benefit from the experiences acquired during the resolution of previous problems through reuse and adaptation mechanisms. Optimization and effective reuse of development methodologies can significantly enhance development productivity and quality (Holdsworth, 1999). Reuse may take place inside a project or across multiple projects, inside an organization or across multiple organizations.

A reusable component is defined as being any design artifact that is specifically developed to be used and is actually used in more than one context (Zhang & Lyytinen, 2001). A large variety of components also called patterns (Fowler, 1997; Gamma, Helm, Johnson, & Vlissides, 1995), business objects (Cauvet & Semmak, 1996), frameworks (Willis, 1996), and COTS or assets (OMG, 2005) have been proposed. Components differ with regard to their granularity, abstraction level (software components, design components, business components), or by the kind of knowledge they allow for reuse (Cauvet & Rosenthal-Sabroux, 2001). Attempts have been made in the field of method engineering to take advantage of reuse from the process point of view. Two research directions have been extensively studied in the field of software component reuse: design for reuse and design by reuse.

Design for reuse aims at proposing systems to support the identification, specification, organization, and implementation of reusable components. Research in this field deals with the definition of models and languages to specify reusable components. These languages allow the modeling of generic knowledge. Associated tools and methods have also been proposed (Cauvet & Rosenthal-Sabroux, 2001).

Design by reuse deals with the development of a new software-based information system by reusing suitable components (Kang, Cohen, Hess, Novak, & Peterson, 1990). Research in this field led to rethinking software-based information system development processes. Dedicated tools to systematically reuse components during the development process have also been proposed: component repositories, reuse systems, and environments of development by reuse (Cauvet, Rieu, Front-Conte, & Ramadour, 2001). Component repositories provide the collection, sometimes organized, of reusable components. Search facilities are based on the internal organization among the components. There is no support for selection, adaptation, and integration activities. Reuse systems focus on component management functionalities in addition to composition, adaptation, and integration support. These tools do not well support the reuse process environment of development by guiding the selection of components and their adaptation. In these systems, software-based information system development is seen as a problem-solving activity; tools provide support during the solving process and mechanisms to ensure adequacy of the proposed solution.

Indeed, with regard to method engineering activities and knowledge, development for reuse is well supported: Proposals have been made to specify, organize, and implement components (method fragments) inside dedicated repositories as it has been discussed. With regard to development by design, attempts have been made in the field of situational method engineering to provide means to adapt and integrate method fragments (Ralyté, 2001), but additional support is still needed. Moreover, few works have dealt with the selection step, that is to say, search facilities and means to provide suitable components (method fragments) during the development process. Indeed, current approaches in situation method engineering have been thought of for method engineers and therefore well cover development for reuse, but there is still a lack of support dedicated to method users, that is to say, support for development by reuse.

In the software development field, existing approaches supporting the search and retrieval of components can be classified into four types (Khayati, 2002; Mili, Valtchev, Di-Sciullo, & Gabrini, 2001):

- Simple keywords and string search
- Faceted classification and retrieval
- Signature matching
- Behavioural matching

Simple keyword searching may result in too many or too few items retrieved or even unrelated items retrieved. The drawback of the faceted-classification search approaches is the difficulty in managing the classification scheme when domain knowledge evolves. Signature matching techniques are dedicated to software components embedding code and are difficult to apply on components providing knowledge about requirements and the way of working, as it is in our case. Behavioural matching techniques are difficult to use when components have complex behaviours or involve side effects. Finally, all these techniques do not provide ways to augment or extend query (Sugumaran & Storey, 2003). Recent approaches focusing on knowledge reuse (more than software component reuse) propose to combine user intention and application domain information to improve support during the selection step (Pujalte & Ramadour, 2004; Sugumaran & Storey).

Avrilioni and Cunin (2001) have proposed the OPSIS approach to effectively reuse process assets. Their approach matches component interfaces with the process parameters and checks the consistency of the resulting process. This approach is based on the concept of view and deals with processes in general and not especially with software-based information system development processes.

Karlson, Agerfalk, and Hjalmarsson (2001) propose a method to adapt software development methodologies by configuring a standard process model. When a project's characteristic matches one of the recurring patterns of project characteristics, then the associated and predefined process configuration is reused. This approach seeks to create reusable process configurations based on experience from earlier projects.

Users and Organization in Method Engineering

Situational method engineering and software reuse fields are rich with a lot of useful works aiming at improving software-based information system development. From our point of view, two dimensions still need some work to be done. Our first concern is with regard to the method user, as it has already been shortly explained previously. Method users are required to understand and control most of the method knowledge used in their organization when they need only a limited extent of it in their daily tasks. Experiments show that they feel development methodologies are too rigid, too prescriptive, and lack support to face the fast evolution of technological and methodological knowledge. Our second concern is with organization-wide method engineering. On one hand, development methodologies show their best in large organizations where they are strongly required to understand, structure, follow, and control the development process, but on the other hand, situational method engineering does not provide answers about how to tailor a development methodology when used as an organization-wide standard. We will discuss more in detail these drawbacks of current approaches in the next subsections.

Method User's Perspective

Current situational method engineering approaches are mostly dedicated to method engineers: They mainly cover the activity of building new and customized development methodologies, which is the task of method engineers. However, method users (i.e., the project member applying or using the methodology) also need to benefit through reuse and adaptation mechanisms from the experiences acquired during the previous software-based information system development activities. All along the development methodology, guidelines are elaborated, refined, and adapted by method users to deal with their daily activities. These guidelines may be useful to other teams and users facing close situations in different projects independent of the functional domain as well as the technical background. Therefore, a third family of proposals about situational method engineering (in addition to the two families presented—approaches aiming at documenting development methodologies through well-defined fragments and approaches aiming at building new development methodologies starting from existing ones) should focus on method fragmentation for method users in order to provide them with guidelines that are reusable while performing their daily tasks (Gnatz, Marshall, Popp, Rausch, & Schwerin, 2001; Karlson et al., 2001; Mirbel & Ralyté, 2006).

Currently, method users are required to know and understand the full development methodology as well as all its concepts to be able to exploit the methodology, most of the time partially. Moreover, guidelines to manage and adopt process models are not detailed enough to support method users through the metaprocess understanding, and method users lack experience and the ability to establish "homegrown" development methodologies or to tailor existing ones (Rossi, Ramesh, Lyytinen, & Tolvanen, 2004). There is a tension between the method in concept (the methodology as formalized in the manual) and the method in action (as interpreted by method users; Fitzgerald, 1997; Lings & Lundell, 2004). Experiments show that it all has negative effects and discredits development methodologies. In the reuse field also, the provided environments are dedicated to experts having strong knowledge about component repositories and the way they are organized. Such repositories are not dedicated to users who are not experts, for whom the major interest of such environments would be assistance in their search for the most suitable components (Pujalte & Ramadour, 2004). Recent approaches combine user intention and application domain information, as well as natural-language techniques (Pujalte & Ramadour) to answer this need. However, they are not easy to deploy and require investment from method users. Moreover, they are based on domain knowledge and therefore are not usable in cross-domain organizations.

Experiments show that method users prefer lightweight development methodologies to heavyweight ones because they feel more implicated. Lightweight development methodologies increase method users' involvement, contrary to heavyweight ones for which the only significant choices are made by method engineers. Feedback from method users shows that development methodologies are seen as too prescriptive and too rigid (Bajec, Vavpotic, & Kirsper, 2004). However, lightweight development methodologies are most of the time empirical processes derived by categorizing observed inputs and outputs, and by defining meaningful controls to keep the process within prescribed bounds (Rising & Janoff, 2000). In empirical development methodology modeling, models are strictly based on experimentally obtained input and output data, with no recourse to any laws concerning the fundamental nature and properties of the system to develop, the project, or the methodology itself. Therefore, it makes it very difficult to transpose it from one organization to another. Moreover, when agile processes, as they are called, increasingly emphasize customization and flexibility, there is no guarantee that a stable process can be established (Rossi et al., 2004).

Increased outsourcing and new development contexts create unforeseen needs for development methodologies management and deployment. Method engineering needs to be analyzed as a continuous, evolutionary process that supports the adaptation of development methodologies to changing technical and organizational contingencies and new development needs. Method rationale and evolutionary method engineering (Rossi et al., 2004) are answers currently provided on this topic. However, they focus on the metamodel evolution, that is to say, the method-engineer work, while we try to provide solutions to support this evolution at the method-user level. Up to now, method fragments have been thought of to support the building of new and better adapted development methodologies, and component reuse has been studied in order to capitalize knowledge about the system to develop. Both of them have been proposed to support method engineers and experts in their work.

New proposals have to focus on method users' needs. They should provide solutions to help method users quickly, easily, and efficiently access method-engineering knowledge. A method-fragment repository should be seen as a repository of experiences about method engineering, and means have to be provided to maintain method-engineering knowledge in a pragmatic-oriented way in order to also focus on method users' activities (in addition to the method engineers' activities). Adaptability means facing the evolution of technologies and development methodologies in the organizations as well as providing development-software-based information system development to both engineers and users.

Organization-Wide Approaches

Works in the field of situational method engineering have dealt with solutions to better handle and answer the need for customization at the project level (Brinkkemper et al., 1998; Harmsen, 1997; Rolland et al., 1998). However, as it has been emphasized in Fitzgerald (1997), little research has focused on how to tailor such situational methodologies when used as organization-wide standard approaches. When situational method engineering allows one to build and customize a standard development methodology into a methodology specific to the organization's need, we call this an organization-wide development methodology. It will usually last for a relatively long period of time because of the commitment and investment it requires. On the other hand, the constant evolution of techniques, mechanisms, and technologies used to develop software-based information systems requires support for methodological evolution, too. Changes in method requirements during the project lifetime evolution have not at all been handled by current approaches in the field of situational method engineering (Agerfalk & Karlsson, 2004). The authors argue that organizations need to explicitly represent the conditions under which various methodology steps are executed in order to enhance the reusability and tailoring of these methodologies.

One solution to deploy an organization-wide development methodology could be to capture and understand each development methodology used inside each project (which we call a project-specific development methodology) to build an organization-wide standard development methodology by merging the best practices coming from all the projects. This solution requires that method engineers are able to capture and understand each variation of each development methodology in each project. It is not an easy task, as it is emphasized in Rossi et al. (2004). It also requires one to make each method user accept and use the new organization-wide development methodology instead of his or her customized version of it. It is also not easy because method users prefer lightweight development methodologies to heavyweight ones for which they feel not enough implicated.

Moreover, evolution would not at all be efficiently supported. Method engineers will have to regularly look at the way the development methodology is applied and the way the technology and the software-based information system evolve to update and maintain the method-engineering knowledge. This solution is not satisfactory, and we propose an alternative organization of method-engineering knowledge into a federation of method fragments as it will be presented in this chapter.

Method-Chunk Federation

In order to overcome the identified shortcomings of current situational method-engineering approaches, we propose an alternative solution that consists of federating the different project-specific development methodologies in order to allow each project to share its method-engineering best practices with the other projects without imposing to all projects a unique organization-wide development methodology. We particularly focus on the needs of method users, which have been discussed in current research only to a limited extent.

We started from the work of J. Ralyté (2001) about method chunks to break down project-specific development methodologies into atomic and reusable parts. Our contribution focuses on the specification and use of a reuse frame to retrieve meaningful method chunks. In our proposal, we provide means to do the following:

- Make the federation possible by introducing a reuse frame in order to capture and share knowledge about software-based information development activities
- Support the federation by providing means for the method users
  - To qualify the content of each atomic and reusable part of the project-specific development methodology through what we call a reuse context
  - To express their need through a user situation

We also propose a similarity metric and a closeness distance to retrieve method chunks, not strictly matching the user need by exploiting the different kinds of refinement relationships that exist in the knowledge-qualifying software-based information system development activities.

Indeed, in our approach, we distinguish (a) the project area where the project-specific development methodology is built and deployed, and (b) the federation where the reusable parts of the project-specific development methodologies are exported from the project area to be shared by all the projects. In our approach, each project starts by breaking its project-specific development methodology into reusable parts (Step 1, method breakdown, in Figure 1). Then, the reusable parts are exported in the federation and qualified with the help of the keywords stored in the reuse frame (Step 2, method-fragment

Figure 1. Overall architecture

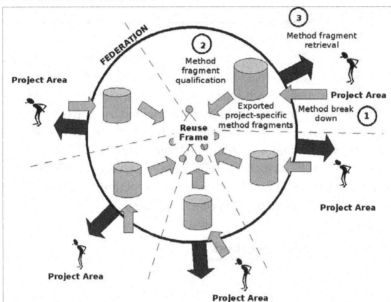

qualification, in Figure 1). Each project exports its reusable parts into a repository that is part of the federation. Qualified method fragments become selectable and able to be queried by the method users from the other projects (Step 3, method-fragment retrieval, in Figure 1).

The rest of the section is organized as follows. First, the notion of method chunk we started with to describe reusable knowledge about software-based information system development activities is described, then we discuss the reuse frame, which allows us to store and organize meaningful knowledge about software-based information system development activities. The method-chunk notion in addition to the reuse-frame structure constitutes the main support to make project-specific development methodologies able to be federated. In a second subsection, we show how method-chunk federation is supported. We start first by introducing the notion of reuse context to qualify method chunks. Then we specify the notion of user situation, which allows users to tell about their methodological needs. Finally we discuss the similarity metrics and the closeness distance we propose to quantify the matching between reuse contexts (when comparing method chunks) or between a user situation and a reuse context.

Making Project-Specific Development Methodologies Able to be Federated

Making project-specific development methodologies able to be federated means to provide ways to break down a development methodology into reusable parts and to qualify each reusable part with meaningful keywords to allow the reusable part to be selected by other method users in other projects. In this section, we first present the model we choose to specify reusable parts of software-based information system development methodologies. Then, we present the *reuse frame,* which aggregates the keywords useful to qualify software-based information system development activities. We discuss the structure of our reuse frame before proposing standard content for it.

Atomic and Reusable Method Part

The notion of reusable method component has been widely studied in the field of situational method engineering, as it has been discussed previously. In our work, we started from the notion of m*ethod chunk* proposed by Ralyté (2001), which is the most complete and suitable reusable asset for our purpose. In Ralyté's approach, a development methodology is viewed as a set of loosely coupled m*ethod chunks* expressed at different levels of granularity. A m*ethod chunk* is an autonomous and coherent part of a development methodology supporting the realization of some specific development activities. Such a modular view permits one to reuse chunks of a given development methodology in others and to federate chunks in our approach. Indeed, the notion of method chunk covers the product and process dimensions, and the map formalism it is based on allows us to define guidelines at different levels of specificity and to support guidelines refinement. It is an intention-based approach of method engineering, very much suitable to support development by reuse. Associated assembly techniques have also been proposed (Ralyté). These operators may be suitable for method users to let them enrich their project-specific development methodology by integrating the new guidelines retrieved in the federation in the project-specific development methodologies.

As part of a development methodology, a m*ethod chunk* ensures a tight coupling of some process part of a development methodology process model and

Figure 2. Method-chunk structure

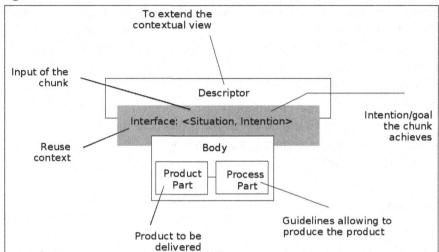

its related product part. The interface of the *method chunk* captures the reuse context in which the *method chunk* can be applied. Besides this, a descriptor is associated with every *method chunk*. It extends the contextual view captured in the chunk interface to define the context in which the chunk can be reused. For more details about the structure and content of a *method chunk*, please refer to Ralyté (2001) and Mirbel and Ralyté (2006). The different elements constituting a method chunk are summarized in Figure 2.

An example of the interface and body part of a method chunk, called business-rule behaviour, is given in Figure 3. This method chunk is extracted from the JECKO methodology developed in collaboration with Amadeus S.A.S. (Mirbel and de Rivieres, 2002).

Figure 3. Method-chunk example

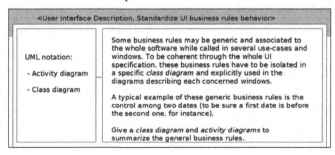

In this method chunk, textual guidelines are given to deal with the description of a user interface and more specifically to help in describing in a standard manner the different business rules related to the user-interface behaviour. For this purpose, the use of UML (unified modeling language) activity and class diagrams is recommended, and textual guidelines are provided. The descriptor of this method chunk will be presented later.

The core elements of our environment are a method-chunk repository to store the method chunks shared by all the projects in the organization and a reuse frame to share a common set of vocabulary to qualify the method chunks of the repository.

The Reuse Frame

The reuse frame is an ontological structure shared by all the method users. It allows structured storage and subsequent retrieval of method chunks. In doing so, the presented approach allows vague retrieval queries and does not rely on a consistent and hard-to-maintain tagging mechanism.

As for the reuse frame, our contribution is twofold. First, we propose a structure allowing one to organize meaningful keywords to qualify method chunks in order to improve their reusability by method users; we also propose standard content for this reuse frame. This content has been elaborated by integrating the different works that have been done to describe contingency factors in software-based information system development projects. It should therefore be suitable for any company. However, it could also be changed or even replaced by other content more suitable for the organization it is deployed in.

In the following, we will start by discussing the reuse-frame structure before proposing our standard content.

Reuse-Frame Structure

The aim of the reuse frame is to capture the knowledge about method engineering that could be discriminant and meaningful to describe method chunks and user needs. It is organized as a tree of criteria that may be described more or less precisely by using refinement relationships. Criteria allow one to qualify reusable method chunks in a simple and more practical way. The level of user involvement and time pressures are examples of discriminant factors (van Slooten & Hodes, 1996) that are helpful to qualify the method-engineering knowledge embedded into a method chunk. Virtual user involvement and real user involvement are examples of specialization of the level-of-user-involvement factor. Indeed, we propose three perspectives to classify criteria in a kind of and-or tree form. The three perspectives we provide to classify criteria are:

- The basic refinement relationship
- The refinement into more specific and classified criteria
- The refinement into more specific and exclusive criteria

The relationship to refine criteria into more specific and classified criteria allows us to specify an order among the nodes sharing the same direct father node. The level of expertise of the method user targeted by the method chunk under qualification is another example of meaningful criteria. Different levels of expertise may be distinguished: expert method users, medium method users, and beginner method users. If a method user searches for method chunks satisfying the expert method users' criteria and no chunks are found, maybe he or she would be interested in looking for method chunks dedicated to medium method users, but not to method chunks dedicated to beginner method users, which would seem unsuitable. Ranking starts from 1 to n (one by one); n is the number of nodes sharing the same direct father node. Therefore, we integrated this kind of refinement relationship into our reuse-frame structure. The refinement into nodes to specify more specific and classified criteria may be helpful when retrieving *method chunks* to find *method chunks* qualified by criteria classified as *previous* or *next to* the criteria of the *method chunk* or of the *user situation* under consideration.

The refinement into nodes to specify exclusive criteria prevents method users from qualifying *method chunks* or *user needs* through incompatible criteria. High time pressure and low time pressure are examples of exclusive refinements of the time-pressure factor introduced previously.

The refinement into nodes to specify more specific criteria may be helpful when retrieving *method chunks* to find *method chunks* qualified by criteria more or *less generic than* the criteria of the *method chunk* or of the *user situation* under consideration.

The different kinds of relationships are summarized in Figure 4.

In the *reuse frame*, nodes close to the root node deal with general criteria while nodes close to leaf nodes (including leaf nodes) deal with precise criteria. A *criterion* is fully defined as a path from the root node to a node *n* of the *reuse frame*. If *n* is not a leaf node, then it should not have exclusive relationships starting from it, otherwise one of the ending nodes of the exclusive relationships has to be chosen as *n*. *Inclusion* between criteria has been defined to state when a criterion is more generic or more specific than another one. A precedence relationship has also been defined to state when a criterion is previous to or next to another one. The *compatibility* between criteria allows one to specify when criteria may be part of the same u*ser situation* or r*euse context*.

Inclusion: A criterion c1 is included in a criterion c2 if the path from the root node to c1 is a subpath of the path from the root node to c2. A criterion c1 includes a criterion c2 if the path from the root node to c2 is a subpath of the path from the root node to c1 .

In Figure 4, N8 *includes* N4.

Precedence: A criterion c1 is *previous* to criterion c2 if they have the same direct father node and the classification rank of c*1* is inferior to the classification rank of c*2*. c1 is *next* to c2 if c1 and c2 have the same direct father node and the classification rank of c*2* is inferior to the classification rank of c*1*.

Figure 4. Reuse-frame refinement relationships

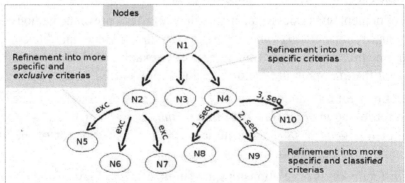

In Figure 4, `N8` is *previous* to N9, and `N10 is *after* N9`.

Compatibility: Included criteria are compatible. If one is not included in the other, criteria are compatible only if they do not share in their path (from the `root` node) a node with `exclusive` leaving edges.

In Figure 4, N5 and N7 are not compatible, while N5 and N3 are compatible.

Reuse-Frame Content: A Proposal

In our approach, method-engineering knowledge is described in terms of criteria, belonging to criteria families, which are successive refinements. In our standard content, we start from the three main dimensions to qualify software-based information system development activities: human, organizational, and technical. Starting from these three basic dimensions, each company may populate the *reuse frame* with its own relevant criteria. We provide *reuse-frame* content that we built from various work made on meaningful criteria for development-methodology characterization (Mirbel & Ralyté, 2006). With regard to the organizational dimension, we started from the work of van Slooten and Hodes (1996), providing elements to characterize software-based information systems development projects: contingency factors, project characteristics, goals and assumptions, as well as system engineering activities. With regard to the technical dimension, we started from previous work on JECKO, a context-driven approach to software development developed in collaboration with the Amadeus Company. In this work, we contribute to the definition of software-critical criteria in order to get suitable documentation to support the software development process (Mirbel & de Rivières, 2002). Our technical dimension also includes criteria related to the source system (as legacy systems are more and more present in organizations) and application technology, which requires more adapted development processes. Finally, regarding the human dimension, we cope with the different kinds of method users that may be involved in the software-based information system development project (analysts, developers, etc.) as well as their expertise level.

Figure 5 shows part of the standard content we propose for the *reuse frame*. In this part of the *reuse frame*, `application type` is an example of *wrong* criteria while `intra-organization application` is an example of a *right* one. `Database` is included in the `application technology`, `medium analyst` is a criterion more specific than `analyst`, `expert analyst` is *next to* `medium analyst` while `beginner analyst` is *previous to* `medium analyst`, and finally `intra-organization application` and `interorganization application` are not compatible criteria while `high complexity` and `high time pressure` are compatible ones. For the full description of our *reuse-frame* content proposal, please refer to Mirbel and Ralyté (2006) and Mirbel (2006).

In this section we presented the method-chunk model and the reuse-frame structure and content. They allow support of the breaking down of project-specific development methodologies into method chunks. In the next section, we will explain how we support method-chunk federation by qualifying method chunks, by qualifying user needs, and by using the similarity metrics and closeness distance to select meaningful method chunks in the federation.

Supporting Method-Chunk Federation

In this section, we present the method-chunk context, aiming at qualifying method chunks built from the different project-specific development methodologies in order to make them retrievable in the federation. Then we discuss the u*ser situation* allowing method users to specify their need and

Figure 5. Part of the reuse-frame content proposal

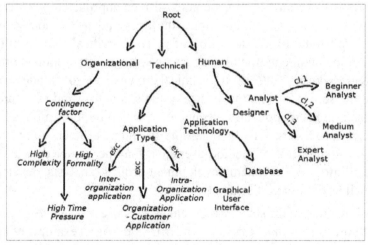

the s*imilarity metrics* we propose to compare method users' needs (i.e., user situation) or project-specific method chunks (i.e., method-chunk context) with the whole set of federated *method chunks*. Then we detail how we extend our s*imilarity metrics* to allow the retrieval of m*ethod chunks* with close *contexts* instead of those that exactly match.

Method-Chunk Qualification

As it has been highlighted before, making development methodology able to be federated means to provide means to break down development methodologies into reusable autonomous and coherent parts and also to provide means to qualify each development-methodology part with meaningful keywords in order to make it retrievable by others. Dedicated efforts have been made in the field of method engineering to provide efficient classification and retrieving techniques to store and retrieve *method fragments*. Classification and retrieving techniques are currently based on structural relationships among fragments (specialization, composition, alternative, etc.) and reuse intention matching. From our point of view, current classification and retrieving means are not fully suitable for a federation of *method chunks* because they are supported by the structure of the development methodology they are a part of. Recent works on method-component reuse combine user intention and application domain information in order to provide alternative and richer means to organize and retrieve components (Pujalte & Ramadour, 2004). But again, domain information does not look like the most suitable information to support a federation as projects may belong to different application domains. The only knowledge that will be understandable by every method user (that is to say, knowledge that is neither application-domain oriented nor project-specific development methodology oriented) is knowledge about method engineering. Therefore, we propose the *reuse frame presented earlier.* Indeed, the descriptor associated with each *method chunk* (which extends the contextual view captured in the chunk interface to define the context in which the chunk can be reused) is specified through a set of at least one criterion taken from the *reuse frame*. It is called the *reuse context* and allows meaningful qualification of *method chunks* in order to allow their reuse through the federation.

The r*euse context* is defined as a set of at least one compatible criterion taken from the *reuse frame*. M*ethod chunks* providing general guidelines are usually associated with general criteria, that is to say, criteria represented by nodes

Figure 6. An example of method chunk reuse context

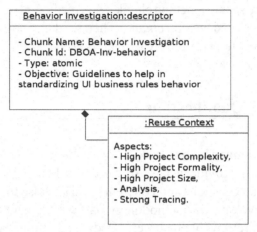

close to the root node. On the contrary, specific guidelines are provided in *method chunks* associated with precise criteria, in other words, criteria corresponding to nodes close to leaf nodes or leaf nodes themselves. It is up to the method engineer who enters the *method chunk* into the federation to select the most meaningful criteria to qualify the *method chunk*. Figure 6 deals again with the method chunk named business-rule behaviour introduced in Figure 3 (body and interface). Figure 6 focuses on the descriptor part of this method chunk. The criteria have been selected among the criteria provided in our reuse-frame content proposal.

User-Need Qualification

The *user situation* allows method users to express their needs to retrieve suitable *method chunks* from the federation. The user situation is specified by a set of criteria selected among those proposed in the reuse frame. In addition to the pertinent criteria, called *necessary criteria*, method users may give *forbidden criteria*, that is to say, criteria he or she is not interested in. It could be helpful in some cases to be sure the *method chunks* including these (forbidden) criteria will not appear in the retrieved set of *method chunks*. All criteria must be compatible among each other inside each set.

If the method user searches for general guidelines, he or she should select necessary criteria that are less refined, that is to say, criteria corresponding to nodes close to the root node of the *reuse frame*. On the contrary, if the method

Figure 7. An example of user situation

: Method User Situation
Necessary Criteria: - Graphical User Interface - Database - High formality - Analyst **Forbidden Criteria:** - Design Eng. Activity - High time pressure

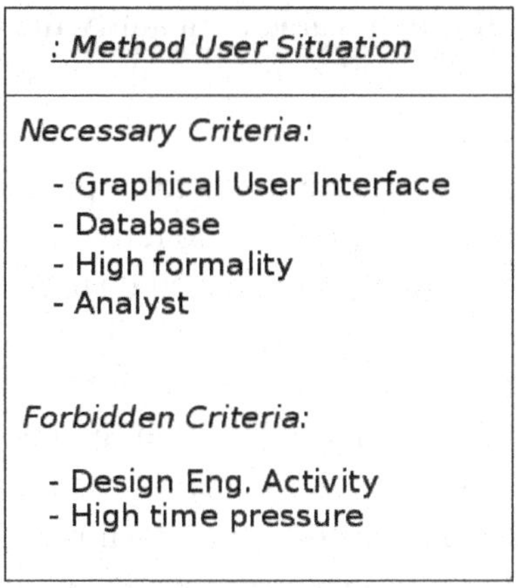

user searches for specific guidelines, he or she may specify the need by selecting criteria that are more refined, in other words, criteria corresponding to nodes close to the leaf nodes or leaf nodes themselves in the *reuse frame*. Figure 7 gives an example of user situation. The criteria have been selected among the criteria provided in our reuse-frame content proposal.

Similarity Metrics and Closeness Distance

Our proposal aims at federating different project-specific development methodologies, allowing each project to share its best practices with the other projects but without imposing to all of them a unique organization-wide development methodology. For this purpose, we provide means for the method user to query the method chunks in the federation to retrieve meaningful method chunks.

A *method chunk* is meaningful (with regard to a method user's need) because it deals with one or several software-based information system development activities covered by the project-specific development methodology the method user usually uses in his or her project; it is therefore an alternative way of working that may be of interest. For this purpose, we defined s*imilarity metrics* to compare a project-specific *method chunk* with the reuse contexts of the method chunks in the federation.

A *method chunk* is also meaningful when it deals with one or several software-based information system development activities that are not (well) covered by the project-specific development methodology. For this purpose, our *similarity metrics* are also applicable to quantify the matching between a user situation and a m*ethod-chunk* context in the set of federated *method chunks*.

The main interest of the federation is the ability to propose new *method chunks* to method users. Means have to be provided to retrieve as many meaningful *method chunks* as possible as an answer to a method user's needs. Therefore, *method chunks* that *reuse contexts that* do not fully match the criteria provided by the method user may also be of interest and have to be shown to the method user. In this case, the similarity between the *user situation* and the *reuse context* has to be quantified. A *reuse context* that does not fully match a *user situation* is, for instance, a *reuse context* whose criteria are included in the *user-situation* list of criteria. The specification of method-engineering knowledge is not something very well defined, and each person making reference to it could understand something slightly different about it. Therefore, guidelines may be more or less detailed in the body of a *method chunk*, and a *method chunk* may be qualified by more or less specific criteria even if shared by all the method users. Therefore, we believe it is meaningful to retrieve *method chunks* qualified by more generic or more specific keywords. Looking at knowledge qualifying software-based information system development activities, one may observe that some of it is ordered. For instance, *expert* designers know more about design than medium ones, who know more than *beginner* ones. Therefore, a *method chunk* dedicated to an expert designer may also be interesting for a medium one, just as a *method chunk* dedicated to a beginner designer may also be interesting for a medium one. Borderlines between ordered criteria (expert, medium, and beginner designers) are not always strictly defined. Therefore, we believe it is meaningful, when retrieving *method chunks,* to search also for *method chunks* associated with criteria *previous* to or *next* to the criteria under consideration in the *user situation*. In this extended kind of retrieval, the similarity between the *user situation* and the *reuse context* of the retrieved *method chunks* has to be quantified. It is the purpose of the similarity metrics.

Similarity Metrics between Method Chunks

In this case, the *reuse context* of two *method chunks* is considered (one from the project-specific *development methodology* and one from the *method-chunk* federation). By looking at the number of common criteria in their *reuse contexts*, a *similarity metric*, sm, varying between 0 and 1, is computed to indicate to the method user how much the *method chunk* from the federation matches the project-specific *method chunk*.

$$sm(mcp,mcf)=[\Sigma^{i=1..n}d(c_{CA_{mcp_i}},CA_{mcf})]/[max(card(CA_{mcp}),card(CA_{mcf})],$$

where *mcp* is the method chunk from the project-specific development methodology and *mcf* is the method chunk from the federation. *CA* denotes the set of criteria of the reuse context, $CA=\{c_1, .., c_i\}$, and $d(c,C)$ is the distance between a criterion c and a set of criteria C defined as follows:

if $c \in C$, *then* $d(c,C) =1$, *else* $d(c,C)=0$.

Similarity Metrics between User Need and Method Chunks

In this case, the retrieval is done by comparing the *reuse context* of the *method chunks* from the federation with a *user situation*. The *similarity metrics* are based on:

- The number of common criteria between the necessary criteria from the *user situation* and the *reuse context*.
- The number of common criteria between the forbidden criteria from the *user situation* and the *reuse context*.
- The number of necessary criteria in the *user situation*.

A positive value of the *similarity metric* indicates that there are more necessary criteria than forbidden ones in the *reuse context* with regard to the *user situation*. On the contrary, a negative value indicates that there are less necessary criteria than forbidden ones. The perfect adequateness is represented by the value 1.

$$sm(rc,us)=[(\Sigma^{i=1..n}d(c_{NC_{us_i}}, CA_{rc}))-(\Sigma^{j=1..m}d(c_{FC_{us_j}}, CA_{rc}))]/[card(NC_{us})],$$

where *us* is the user *situation*, $NC_{us} = \{Nc_{us}1, .., NC_{us}i\}$ is its necessary criteria set, and $FC_{us} = \{FC_{us}1, .., FC_{us}j\}$ is its forbidden criteria set; *rc* a *reuse context*, CA_{rc} is its set of criteria, and *d(c'C)* is the distance between a criterion *c* and a set of criteria *C* defined as follows:

if c ∈ C, then d(c,C)=1, else d(c,C)=0.

Examples of *reuse contexts* and *user situations* are given in Figure 8. Similarity metrics have been computed. The example shows that the two *method chunks* under consideration better match the first *user situation* than the second one. The first *method chunk* fully matches the *user situation* A.

Extended Similarity Metrics

Method chunks including more general, more specific, previous, or next criteria in their *reuse context*, with regard to the criteria of the *reuse situation*, are also of interest, as described previously. They are considered *close* method chunks. Method chunks including more specific criteria in their *reuse context* may be of interest because they usually provide more specific guidelines; they may better cover part of the methodological problem the

Figure 8. Examples of similarity metrics between user situation and reuse context

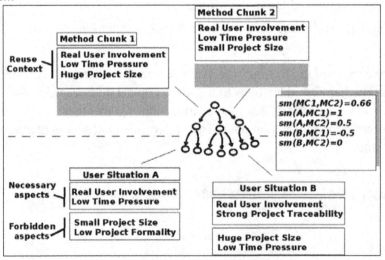

method user is interested in. If one searches, for instance, for method chunks matching `code-reuse` criteria, he or she may also be interested in method chunks matching the `weak code reuse`, `medium code reuse`, and `strong code reuse` as shown in Figure 9. The reuse-frame content used for this example is taken from our standard content proposal. A method user may also be interested in method chunks qualified by more general criteria because they may provide more general-purpose guidelines that could also be of interest. In the same way, the classification feature of refinement relationships may be exploited to enlarge the set of retrieved method chunks with *method chunks* that *reuse contexts* including previous and next criteria. Examples are shown in Figure 9.

Exploiting *reuse-frame* refinement relationships may also be interesting with regard to *forbidden* criteria. Indeed, enlarging the set of forbidden criteria to more general ones means to forbid full branches of the *reuse frame,* and enlarging the set of forbidden criteria to more specific criteria means to forbid *method chunks* associated with too-specific criteria, most probably qualifying *method chunks* providing too-specific guidelines. In the same way, enlarging the set of forbidden criteria with criteria *previous* to or *next to* the criteria

Figure 9. Example of extension though refinement relationships

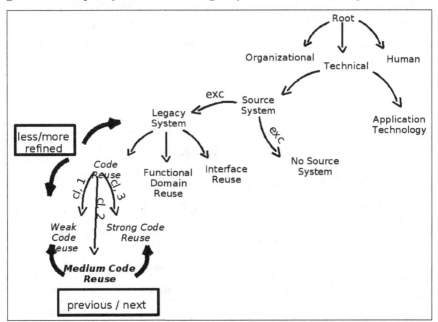

under consideration (through classified refinement relationship) means to avoid retrieving *method chunks* whose scopes overcome the criteria given by the method user.

Extending the selection by allowing or not allowing more general, more specific, previous, or next criteria to be included in the necessary and/or forbidden criteria given in the *user situation* provides a way for the method user to reduce or enlarge the number of *method chunks* retrieved. If a method user does not find enough *method chunks* with regard to the methodological need (i.e., the user situation), the method user may choose to use the extended similarity metrics (in this way taking into account more general, more specific, previous, and/or next criteria) in order to find more *method chunks*. On the contrary, if the set of *method chunks* provided as an answer to the method user is too large, the set of forbidden criteria may be enlarged by applying the extended similarity metrics to the forbidden criteria (they may take into account more general, more specific, previous, and/or next criteria) and in this way reduce the number of retrieved *method chunks*. The table presented in Figure 10 summarizes the extension possibilities.

When the *similarity metrics* are computed with extended necessary and/or forbidden criteria, a distance has to be provided to quantify the closeness between the criteria under consideration in the user situation or reuse context and the more generic, more specific, previous, or next criteria in the reuse context of the method chunk in the federation. Therefore, we propose a definition for what we call the extension of a node (that is to say, the set of criteria more generic, more specific, previous to, or next to it) and a closeness distance to qualify the closeness between criteria.

Figure 10. Similarity metrics and extension possibilities

	Exact Matching	Extended Matching		
		Less refined aspects	More refined aspects	previous/next aspects
Necessary aspects	to search for method chunks	to retrieve *more* method chunks		
		More general chunks	More specific chunks	overlapping method chunks
Forbidden aspects	to avoid method chunks	to retrieve *less* method chunks		
		to avoid too general chunks	to avoid too specific chunks	to avoid overlapping chunks

Extension: The extension *ext(a)* of a criterion a is the set of all the criteria more generic, more specific, previous to it, and next to it.

In Figure 5, the extension of Application Technology, for instance, is {Technical, Application Technology, Graphical User Interface, Database}. A closeness distance is proposed to qualify the closeness between two criteria. A perfect matching between two criteria leads to the value 1 of this *closeness distance*, which moves toward 0 as far as the ratio decreases. The root node gets the value 0 if compared to a criterion in the reuse frame. With regard to more specific criteria, the value 0 is associated with the leaf node of the deepest

Figure 11. Closeness distance boundaries

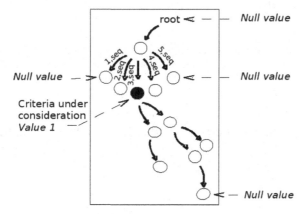

Figure 12. Example of closeness distance with generic criteria

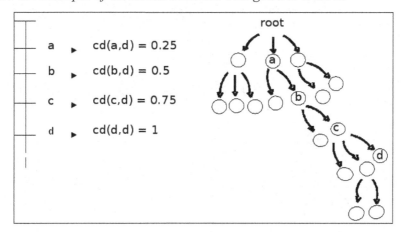

branch of the subtree rooted by the criteria under consideration. With regard to previous and next criteria, the value 0 is associated with the nodes that rank the lowest and the highest among the nodes sharing the same direct father node as the criteria under consideration. Figure 11 summarizes all the cases of null value.

Examples of closeness distances are given in Figure 12. In this example, the criterion d is extended by $\{a,b,c,d\}$. A closeness distance is computed for each criterion belonging to the extension.

The *similarity metric* is expandable thanks to this *closeness distance*. When the criteria that are present in the *reuse context* of the *method chunk* from the federation under consideration do not strictly match the criteria of the *user situation or the criteria of the project-specific method-chunk reuse context*, their extensions are computed and the *closeness distance* is used in the computation of the *similarity metric*.

$$sm(rc,us)=[(\Sigma^{i=1..n}\,ed(c_{NC_{us_i}},\,CA_{rc}))-(\Sigma^{j=1..m}\,ed(c_{FC_{us_j}},\,CA_{rc}))]/[card(NC_{us})],$$

where *us* is the user *situation*, $NC_{us} = \{NC_{us}1\,,\,..,\,NC_{us}i\,\}$ is its necessary criteria set, $FC_{us} = \{FC_{us}1,\,..,\,FC_{us}j\}$ is its forbidden criteria set, *rc* is a *reuse context*, CA_{rc} is its set of criteria, and $ed(c_1C)$ is the extended closeness distance.

Figure 13. Example of extended similarity metrics

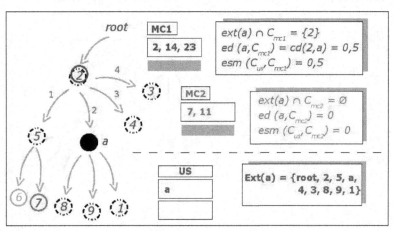

Examples of extended similarity metrics are shown in Figure 13. A simple user situation characterized by the singleton a is compared to the reuse contexts of method chunks MC1 and MC2. There is no intersection between the extension of a and the MC2 criteria, so the extended similarity metric is therefore null. There is one criterion common to the extension of a and MC1's context, so the closeness distance is computed for Criterion 2 and the extended similarity metric is also computed.

Future Trends

In this chapter we presented an approach to federate different deployments of development methodologies belonging to the same organization without imposing a unique organization-wide development methodology. Our contribution aims toward a twofold goal:

- To better dedicate method-engineering environments to method users (nonexpert users) to allow them to take advantage of previous best practices about method engineering as well as what method engineers do in current situational method-engineering proposals (to perform their work, which consists of building and deploying adequate development methodologies in the organization).

- To better support situational method-engineering deployment at the organization level, especially in organizations built by merging smaller organizations: In this kind of context, each of the small organizations comes with its own projects and specific development methodologies. There is a lack of support to harmonize all the development processes by imposing a new and unique organization-wide development methodology.

With regard to the effort provided to better support method users during method engineering by reuse, our work focused on the selection phase by providing means to search for method fragments and means to qualify the adequacy between the method fragments and the user needs. Support on the step of the method engineering by reuse dealing with the adaptation and the integration of the retrieved method fragment(s) to the project-specific development methodology still has to be proposed. Means to take into account method-user feedback about the way method fragments have been reused still have to be studied more in detail, too.

With regard to organization-wide methodological support, the management of local versions (or configurations) of development methodologies at the project level and their alignment with the centralized method-fragment repository have to be better investigated. Refinement and versioning means also have to be proposed to enhance the usability of a method-fragment repository in the context of a federation of development processes.

Indeed, our work falls under the line of research conducted in the field of situational method engineering. In this field, a lot of work has been done to provide means to capture project specificities and to customize methodologies to fit these specificities. However, this working direction has left aside the need for lightweight methods claimed by method users. We believe method engineering now has to move from a directed way of presenting the development methodologies into a more federated way where each individual project inside the organization may contribute to the shared development methodology without having it imposed and deployed in a unique way. Future method-engineering environments should tend to support communities of method users, putting in common their method-engineering knowledge. Profiles should be specified at the individual and/or the project levels. This would allow the comparing of individuals and projects. Inside the organization, it would allow the discovery of a community of users and similar projects (from the methodological point of view). It would allow commonly agreed-on best practices to emerge at the organization, at the projects or individuals levels. Profiles would also help to better quantify method-fragment reusability with regard to groups of projects or groups of users. One step further would be to propose a framework to support and enhance collaborative method engineering.

Conclusion

In this chapter, we presented an approach to federate project-specific development methodologies in order to allow each project to capitalize and share its methodological best practices with the other projects of an organization without imposing to all of them a unique organization-wide development methodology.

Our work takes its root in the domain of situational engineering and software reuse. Our contribution aim was twofold. We particularly focused on the needs and requirements of method users (i.e., nonexperts). We proposed means to reassemble the slightly different project-specific methodologies into a shared federation of method-engineering best practices, in this way enhancing the usability of situational method-engineering approaches when used as organization-wide standards. Both points have been discussed in current research only to a limited extent.

In this chapter, we first explained how to make the federation possible by introducing the two core elements of our environment: the method-chunk repository and the reuse frame.

The method-chunk repository consists of a set of atomic and reusable parts of development methodologies that we call method chunks, following the definition given by Ralyté (2001). Method chunks are identified and extracted from the project-specific development methodologies to be exported into the repositories of the federation.

The reuse frame consists of criteria organized in an and-or tree structure. It supports the structured storage and subsequent retrieval of method fragments. Thanks to this reuse frame, our approach *does not rely on a* consistent and hard-to-maintain tagging mechanism. In doing so, it will be much easier for nonexperts to reuse knowledge about method engineering.

Then we explained how we support method-chunk federation by providing the following:

- **Means for method engineers to qualify method chunks through a reuse context:** This reuse context consists of a set of criteria taken from the reuse frame.

- **Means for method users to express their need through a user situation:** A user situation consists of a set of necessary criteria to specify the user need and an optional set of criteria to better indicate the method-engineering knowledge the method user is not interested in.

A similarity metric was also provided to compare the following:

- Method-chunk reuse contexts between them to find similar method chunks
- Reuse contexts and user situations to find suitable method chunks with regard to a methodological need

Thanks to the different kinds of refinement relationships provided in the reuse-frame structure (basic, classified, or exclusive) and the different kinds of information provided in the user situation (necessary and forbidden criteria), we have shown how the method-chunk retrieval process is tunable and how vague retrieval queries are possible. A closeness distance has been discussed to quantify this vagueness.

The work that has been presented in this chapter is a first step on the way to communities of method users. Our goal is now to move to an environment for collaborative method engineering. It requires that we concentrate our efforts not only on method-engineering knowledge and communication management, but also on means to fully support the development processes in a collaborative way.

Acknowledgment

We would like to thank Jolita Ralyté, Michel Léonard, and Jean-Louis Cavarero for their many useful comments about this work.

References

Agerfalk, P. J., & Karlsson, F. (2004). Method configuration: Adapting to situational characteristics while creating reusable assets. *Information and Software Technology, 46*(9), 619-633.

Avrilioni, D., & Cunin, P. Y. (2001). Process model reuse support: The OPSIS approach. Presented at the 10th International Software Process Workshop.

Bajec, M., Vavpotic, D., & Kirsper, M. (2004). *The scenario and tool-support for constructing flexible, people-focused system development methodolo-*

gies. Presented at the International Conference on Information System Development.

Benjamin, V., & Fensel, D. (1998). Editorial: Problem-solving methods. *International Journal of Human-Computer Studies, 49*, 305-313.

Brinkkemper, S., Saeki, M., & Harmsen, F. (1998). *Assembly techniques for method engineering*. Presented at the International Conference on Advanced Information Systems Engineering.

Cauvet, C., Rieu, D., Front-Conte, A., & Ramadour, P. (2001). Réutilisation dans l'ingénierie des système d'information. In C. Cauvet & C. Rosenthal-Sabroux (Eds.), *Ingénierie des systèmes d'information*. Hermès, France.

Cauvet, C., & Rosenthal-Sabroux, C. (2001). *Ingénierie des systèmes d'information*. Hermès, France.

Cauvet, C., & Semmak, F. (1996). *Semantic units and connectors: Towards domain knowledge reuse*. Presented at the IFIPWG8 Conference on Domain Knowledge for Interactive Systems Design.

Cossentino, M., & Seidita, V. (2004). Composition of a new process to meet agile needs using method engineering. *Software Engineering for Multi-Agent Systems, 36*-51.

Deneckère, R., & Souveyet, C. (1998). *Patterns for extending an OO model with temporal features*. Presented at the International Conference on Object-Oriented Information Systems.

Firesmith, D. G., & Henderson-Sellers, B. (2002). *The OPEN process framework: An introduction*. Harlow, UK: Addison-Wesley.

Fitzgerald, B. (1997). The use of systems development methodologies in practice: A field study. *The Information Systems Journal, 7*(3), 201-212.

Fowler, M. (1997). *Analysis patterns: Reusable object models*. Addison-Wesley.

Gamma, E., Helm, R., Johnson, R., & Vlissides, J. M. (1995). *Design patterns: Elements of reusable object-oriented software*. Addison-Wesley.

Ginsberg, M., & Quinn, L. (1995). *Process tailoring and the software capability maturity model (CMU/SEI-94-TR-024)*. Software Engineering Institute.

Gnatz, M., Marshall, F., Popp, G., Rausch, A., & Schwerin, W. (2002). *Towards a tool support for a living software development process*. Presented at the 35th Hawaii International Conference on System Sciences.

Graham, I., Henderson-Sellers, B., & Younessi, H. (1997). *The OPEN process specification.* Harlow, UK: Addison-Wesley.

Harmsen, A. F. (1997). *Situational method engineering.* Utrecht, the Netherlands: Moret Ernst Young.

Harmsen, F., Brinkkemper, S., & Han Oei, J. L. (1994). *Situational method engineering for informational system project approaches.* Presented at the IFIP WG8.1 Working Conference on Methods and Associated Tools for the Information Systems Life Cycle.

Holdsworth, J. (1999). *Software process design: Out of the tar pit.* London: McGraw-Hill International (UK) Limited.

Kang, K., Cohen, S., Hess, J., Novak, W., & Peterson, S. (1990). *Feature-oriented domain analysis (FODA) feasibility study* (CU/SEI-90-TR-21). Pittsburgh, PA: Software Engineering Institute, Carnegie-Mellon University.

Karlson, F., Agerfalk, P. J., & Hjalmarsson, A. (2001). *Method configuration with development tracks and generic project types.* Presented at the CAISE/IFIP8.1 International Workshop on Evaluation of Modeling Methods in System Analysis and Design (EMMSAD'01).

Khayati, O. (2002). *Components retrieval systems.* Presented at the OOIS Workshop on Reuse in Object-Oriented Information Systems Design.

Leppänen, M. (2006). Towards an ontology for information systems development. In *Proceedings of the 9ᵗʰ International Workshop on Exploring Modeling Methods in Systems Analysis and Design* (pp. 363-374).

Lings, B., & Lundell, B. (2004). *Method-in-action and method-in-tool: Some implications for CASE.* Presented at the 6ᵗʰ International Conference on Enterprise Information Systems.

Mili, H., Valtchev, P., Di-Sciullo, A., & Gabrini, P. (2001). Automating the indexing and retrieval of reusable software components. In *Proceedings of the 6ᵗʰ International Workshop NLDB* (pp. 75-86).

Mirbel, I. (2004). *Rethinking information system development methods: Fitting project team members profiles.* Presented at the International Conference on Information System Development.

Mirbel, I. (2006). *The reuse frame.* Retrieved from http://www.i3s.unice.fr/~mirbel/reuse-frame/html/rf.html

Mirbel, I., & de Rivières, V. (2002). Adapting analysis and design to software context: The JECKO approach. In *Proceedings of the International*

Conference on Object-Oriented Information Systems (OOIS'02) (pp. 223-228).

Mirbel, I., & Ralyté, J. (2006). Situational method engineering: Combining assembly-based and roadmap-driven approaches. *Requirement Engineering Journal, 11*(1), 58-78.

OMG. (2005). *Reusable assets specification.* Retrieved from http://www.omg.org/

Prakash, N. (1999). On method statics and dynamics. *Information Systems, 34*(8), 613-637.

Pujalte, V., & Ramadour, P. (2004). Réutilisation de composants: Un processus interactif de recherche. *Majecstic'05.*

Punter, P., & Lemmen, K. (1996). The MEMA model: Towards a new approach for method engineering. *Information and Software Technology, 38*(4), 295-305.

Ralyté, J. (2001). *Ingénierie des méthodes à base de composants.* Unpublished doctoral dissertation, University of Paris-Sorbonne.

Ralyté, J., & Rolland, C. (2001). An assembly process model for method engineering. In *Proceedings of the 13th International Conference on Advanced Information Systems Engineering (CAISE'01)* (pp. 267-283).

Rising, L., & Janoff, N. S. (2000). The Scrum software development process for small teams. *IEEE Software, 17*(4), 26-32.

Rolland, C., Nurcan, S., & Grosz, G. (2000). A decision making pattern for guiding the enterprise knowledge development process. *Journal of Information and Software Technology, 42*, 313-331.

Rolland, C., & Plihon, V. (1996). *Using generic chunks to generate process models fragments.* Presented at the IEEE International Conference on Requirements Engineering (ICRE'96).

Rolland, C., Plihon, V., & Ralyté, J. (1998). Specifying the reuse context of scenario method chunks. In *Proceedings of the 10th International Conference on Advanced Information System Engineering (CAISE'98)* (pp. 191-218).

Rossi, M., Ramesh, B., Lyytinen, K., & Tolvanen, J. (2004). Managing evolutionary method engineering by method rationale. *Journal of the Association for Information Systems, 5*(9), 356-391.

Storrle, H. (2001). Describing process patterns with UML. *8th European Workshop on Software Process Technology*, 173-181.

Sugumaran, V., & Storey, V. C. (2003). A semantic-based approach to component retrieval. *The Database for Advances in Information Systems, 34*(3).

Sutcliffe, A. G., & Maiden, N. A. M. (1992). Supporting component matching for software reuse. In *Proceedings of the International Conference on Advanced Information Systems Engineering* (pp. 290-303).

Van Slooten, K., & Hodes, B. (1996). Characterizing IS development projects. *IFIP WG 8.1 Conference on Method Engineering*, 29-44.

Willis, A. C. (1996). *Frameworks and component-based development.* Presented at the International Conference on Object-Oriented Information Systems.

Zhang, Z., & Lyytinen, K. (2001). A framework for component reuse in a metamodelling-based software development. *Requirement Engineering Journal, 6*(2), 116-131.

Chapter VII

Modeling and Analyzing Perspectives to Support Knowledge Management

Jian Cai, Peking University, China

Abstract

This chapter introduces a generic modeling approach that explicitly represents the perspectives of stakeholders and their evolution traversing a collaborative process. This approach provides a mechanism to analytically identify the interdependencies among stakeholders and to detect conflicts and reveal their intricate causes and effects. Collaboration is thus improved through efficient knowledge management. This chapter also describes a Web-based information system that uses the perspective model and the social network analysis methodology to support knowledge management within collaboration.

Introduction

The ability to effectively manage distributed knowledge and business processes is becoming an essential core competence of today's organizations. Various knowledge management theories and approaches have been proposed and adopted (Earl, 2001). These include ways to align knowledge processes with strategies (Spender, 1996), to leverage organizational learning abilities (Nonaka & Takeuchi, 1995), and to build IT infrastructures to support knowledge activities (Lu, 2000; Zack, 1999). Knowledge management systems (KMSs) can be viewed as the implementation of the KM strategy. KMS improves the knowledge processes through IT infrastructures and information-processing methodologies (Tanriverdi, 2005). Although the importance of knowledge management has been well recognized, organizations are still facing the problems of how to successfully implement knowledge management. In order to effectively utilize these theories and technologies to support teamwork, it is necessary to gain more fundamental understandings of the characteristics of knowledge management within collaboration processes.

Background

Previous knowledge management approaches can be generally classified into two categories (Hanson, Nohira, & Tierney, 1999). The strategies supporting knowledge replication provide high-quality, fast, and reliable information systems implementation by reusing codified knowledge. The strategies supporting knowledge customization provide creative, analytically rigorous advice on high-level strategic problems by channeling individual expertise. The codification approaches view information technology as the central infrastructure of knowledge-based organizations. KMSs are thus treated as system-integration solutions or applications that retain employees' know-how. The major concern of these approaches is how to help organizations monitor the trends of rapidly changing technologies and inventions in order to recognize new applications that may provide competitive advantage (Kwan & Balasubramanian, 2003). However, IT is just one of the elements of KMS. As knowledge management involves various social and technical enablers, the scope, nature, and purpose of KMS vary during the collaboration processes. Researches from the knowledge-customization perspective focus on understanding knowledge and its relationships with organizations (Becerra-Fernanaez & Sabherwal, 2001; Nonaka & Takeuchi, 1995). A typology of knowledge creation and conversion of tacit and explicit knowledge was proposed (Nonaka, Reinmoeller, & Senoo, 1998). The conversion involves transcending the self of individuals, teams, or organizations and reveals the importance of organizational architecture and organizational dynamics to capitalize on knowledge. Recent research on knowledge management has been focusing on developing models that interconnect knowledge management factors, such as collaboration, learning, organizational structure, process, and IT support (Lee & Choi, 2003). These research works have been mainly addressing understanding the nature of knowledge and knowledge management. Both approaches provide workable models and methods for implementing knowledge management.

In fact, knowledge replication is interlaced with knowledge customization within a collaborative process. In collaborative projects, it is important to systematically integrate these two groups of KM approaches to build methodologies and systems to facilitate the teamwork. First, KM methodologies should be coupled with process management in collaborative projects. An organization and its members can be involved in multiple knowledge management process chains. The tangible tasks are accompanied by the implicit knowledge-integration activities. As such, knowledge management is not a monolithic but a dynamic and continuous organizational phenomenon (Alavi & Leidner, 2001). Second, KM and KMS have to take account of various social factors within collaboration processes. Collaborative projects involve various stakeholders (i.e., all of the human participants and organizations who influence the collaboration process and the results) from different disciplines to work cooperatively over distance and time boundaries. When many heterogeneous groups work together on large projects over a long period of time, their knowledge of the system, the product, and other people will keep on evolving (Dym & Levitt, 1991; O'Leary, 1998). The professional expertise in particular is framed by a person's conceptualization of multiple, ongoing activities, which are essentially identities, comprising intentions, norms, and choreographies (Carley & Prietula, 1994; Erickson & Kellogg, 2000; Siau, 1999; Sowa & Zachman, 1992). Although the collaboration process might appear relatively technical, it is essentially a social construction process when different persons perform their tasks within various adaptive situations (Berger & Luckman, 1966; Clancey, 1993, 1997). The situations will eventually impact the evolution of participants' roles and form a shared understanding (Arias, Eden, Fischer, Gorman, & Scharff, 2000). Even within well-defined technical roles, every stakeholder makes the role his or her own by adapting or executing the role based on his or her conceptions and circumstances. It is the social interaction that determines the variation or adaptability of these roles in a particular application context. As their roles evolve, stakeholders' learning customs and attitudes will vary, which will directly or indirectly affect their internal knowledge and the knowledge creation and conversion processes. Therefore, to manage the distributed knowledge within the complicated collaborative process, it is necessary to have well-developed methodologies for describing and analyzing the social interactions in collaborative contexts of the emerging practice.

This chapter presents a methodology for supporting knowledge management within collaboration by modeling and analyzing the stakeholders' perspectives. The methods to depict and control the evolution of distributed knowledge are introduced. This chapter also describes a prototype knowledge management system developed for a U.S. government research institute. It implements the methodology and uses the advanced network computing techniques to facilitate stakeholders' interaction within their work practice.

Modeling Perspectives
to Support Knowledge Management

The previous approaches and methodologies for supporting KM in collaborative work have been mainly concentrating on either modeling the explicit knowledge or supporting communication of implicit knowledge. The knowledge management systems built upon these approaches included three types of functions: (a) the coding and sharing of best practices, (b) the creation of corporate knowledge directories, and (c) the creation of knowledge networks (Alavi & Leidner, 2001). Recent research has proposed systems to support information and knowledge seeking and use within the decision or problem-solving process (Kwan & Balasubramanian, 2003; Rouse, 2002; Shaw, Ackermann, & Eden, 2003). Modeling approaches are widely used for developing such methodologies and systems. For instance, the activity modeling approach was used to develop a knowledge management system to provide a computer-based guidance and interactive support for office workers (Reimer, Margelisch, & Staudt, 2000). Knowledge-engineering processes were modeled to capture, store, and deploy company knowledge (Preece, Flett, & Sleeman, 2001). However, most of the existing approaches still view stakeholders as homogeneous and do not emphasize their intricate needs at various stages of the processes. Nevertheless, a lack of understanding of stakeholders' needs—and the provision of support systems accordingly—is precisely the missing link in the success of many information and knowledge management systems (Rouse). This requires understanding multiple aspects of stakeholders' needs in seeking and using information and knowledge within the collaboration.

Recent published studies have shown that besides technologies, the social aspects are essential to the success of collaboration (Briggs, Vreede, & Nunamaker, 2003; Easley, Sarv, & Crant, 2003; Erickson & Kellogg, 2000; Hardjono & van Marrewijk, 2001). Technologies are part of a social network and a KM system is likely to include not only technology, but also social and cultural infrastructures and human agents (Chae, Koch, Paradice, & Huy, 2005). One of the key social factors is the cognitive interaction process. As stakeholders' preferences, environments, and knowledge are dynamically changing during their interactions, collaborative activity over the Internet is more than an online data-accessing and information-sharing process. Accordingly, frequently occurred conflicts influence the project schedule and team performance. Team coordination has to be achieved through not only the sharing of data and information, but also the realization of the decision contexts of each other (Chung, Kim, & Dao, 1999; Kannapan & Taylor, 1994). The decision context consists of at least two parts: the circumstances of the decision makers and the stages of the process. When people exchange information, they should understand under what circumstances this information is generated and in which situation it can be potentially used. Otherwise, it is difficult for them to interpret the purposes and implications of each other during the activity coordination. Therefore, to represent and organize the situated knowledge (i.e., the context) is essential to support the coordination among different groups. It is also of immense importance to understand how to design knowledge management systems so that they mesh with human behavior at the individual and collective levels. By allowing users to "see" one another and to make inferences about the activities of others, online collaboration platforms can become environments in which new social forms can be invented, adopted, adapted, and propagated—eventually supporting the same sort of social innovation and diversity that can be observed in physically based cultures (Erickson & Kellogg, 2000).

Figure 1. The perspective modeling approach of knowledge management in collaboration

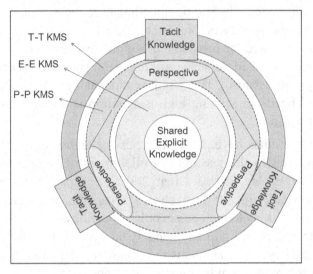

To address these issues, our research uses a sociotechnical framework to model the interactions within collaborations (Lu & Cai, 2001). The framework addresses that one cannot utilize information to map from "what to do" to "how to do" in the collaboration process without knowing the perspective of the "who" that generates the information. A collaborative project is modeled as a coconstruction process among a group of stakeholders. The key feature is to explicitly model the who (i.e., the stakeholders' perspectives) within the process (i.e., the what, how, and when). During collaboration, each individual has a perspective that evolves over time and acts like a lens through which she or he understands and collects information external to her or him. Each individual builds over a lifetime an evolving base of information that is internal. The information that each individual produces, or exchanges through any medium (e.g., computers, speech, and writing), is the external manifestation of internal information, appropriately filtered through "perspective lens." Based on the sociotechnical framework, knowledge management systems require the explicit modeling of stakeholders' perspectives within their social interactions. The perspective modeling and analyzing methodology focuses on representing and handling the interactions among the heterogeneous stakeholders. It provides associations with other knowledge management and decision-support models. It also provides ways to build and integrate various processes with the realization of sharing knowledge and managing conflict. Different from traditional KMS, which either focuses on the codification of explicit knowledge (E-E KMS) or communication of tacit knowledge (T-T KMS), the perspective modeling approach will realize a new way of building KMS (P-P KMS) through controlling the interfaces between the explicit and tacit knowledge (i.e., stakeholders' perspectives; Figure 1).

Perspective Modeling and Sociotechnical Analysis

Methodology Overview

The central function of the research framework is the sociotechnical analysis to model and analyze the perspectives of stakeholders at each step of the collaboration process. The sociotechnical analysis methodology takes three input parameters (i.e., the concept model, the perspective model, and the process model; Figure 2). The concept model is a structure that organizes the ontology models representing the shared or private notions of the stakeholders. The process model is a feasible computational model that represents the interactions of individual tasks. It specifies the sequences and dependencies of decisions and actions to be jointly performed. The perspective model provides a generic means to formally capture, represent, and analyze stakeholders' perspectives and their interactions with each other. The concept model and perspective models represent the shared knowledge and social characteristics of various stakeholders during the collaboration process. They are derived from the surveys of stakeholders' attitudes toward the ontology models at a point of time.

The dependencies among these models are represented as matrices for mathematical analysis. Conflict analysis applies systematic strategies to analyze inconsistencies among these matrices. At a certain stage within the process, conflicts can be detected by tracking and comparing the perspective states of different stakeholders associated with a certain task. This analysis will derive three major outputs (i.e., process feasibility, conflict possibility, and perspective network). Then, based on these outputs, the systems can apply various control strategies so that the quality of the collaboration is enhanced. Control mechanisms adaptively handle the interplay among the three factors

Figure 2. The sociotechnical analysis methodology for knowledge management

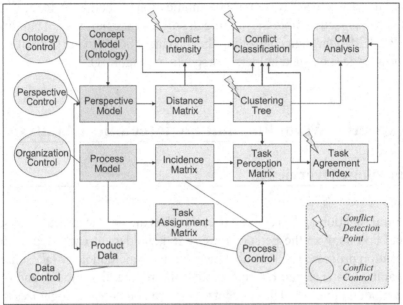

by systematically reconciling various perspectives, improving the processes, and controlling the product data and organizational structure.

Perspective Modeling

The perspective modeling mainly consists of building the concept model and the perspective model. While the process model depicts the tangible activities of the project, the concept model and perspective model track the knowledge evolution and changes of social behaviors.

The first step is to generate the concept structure hierarchy. A concept model is a hierarchical structure that represents the organization of the ontology (Huhns & Stephens, 1999; Staab, Schnurr, Studer, & Sure, 2001) that stakeholders propose and use in their collaboration. Figure 3 shows a concept structure example of a product development team. Stakeholders may use both top-down and bottom-up construction methods (Vet & Mars, 1999) to build the concept structure. It is possible to apply some templates (e.g., product function template, organizational template, conflict types template, etc.) to clarify the concepts. These templates act as the contend-based skeletons for organizing the external information that stakeholders may share with others.

When stakeholders propose new concepts, the concept structure is updated and is used to systematically organize these concepts and their relationships. Since a stakeholder should first consider whether there are same or similar concepts in the structure, only the novel concepts can be specified and added. The concepts involved within the collaboration are classified into two types. Shared concepts are those that have been well defined from previous projects. They have widely accepted meaning shared among the stakeholders (e.g., in Figure 3, Function Requirements, Product, and Organization are shared concepts). Private concepts are perceived only by some particular stakeholders. Their names or meanings are not expressed around the group. If a group of people have a shared purpose toward a concept, everyone will be asked

Figure 3. A concept structure built by stakeholders in a collaborative design project

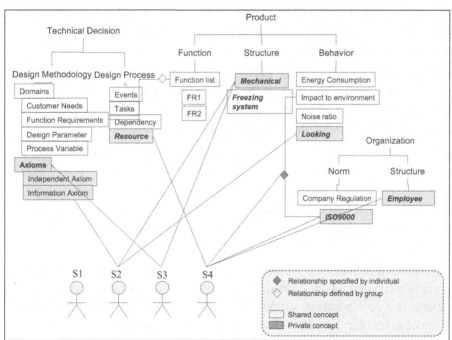

to view it. After the concepts are identified, the dependencies among these concepts can be further clarified by stakeholders.

The second step is to generate the perspective model. A perspective model is the special information representing the status of a stakeholder's perspective at a certain time. A perspective model consists of the purpose (i.e., the intention to conduct certain actions), context (i.e., the circumstances in which one's action occurs), and content (i.e., what one knows and understands) that the stakeholder uses to access the external knowledge and to expose the internal knowledge. In information systems, the perspective model can be depicted as a data format relating to other information entities.

Our research develops a format for representing perspectives and a procedure to capture, generate, and analyze perspective models. Given the well-organized structure of concepts, it is feasible to ask the stakeholders to build the perspective-model state diagrams (PMSDs) at a certain time. A stakeholder's PMSD attempts to depict the explicit relationships among his or her concepts (including the shared concepts and private concepts) and purpose, content, and context information. The concepts listed in the PMSD are categories of perspective contents. Using the concept structure to generate the PMSD provides a structured way for us to systematically compare and examine the perspective differences among stakeholders.

Each concept of the concept model can be associated with a stakeholder by a set of purposes, contexts, and contents. The operation is to ask the stakeholders to do the following.

First, relate this concept to their purposes. A stakeholder is able to specify his or her purpose within the project for a given concept. There might be more than one purpose involved. For an abstract concept, the purpose could be more general. For a specific concept, the purpose could be detail.

Second, specify the relationships of this concept with other concepts based on his or her context. If there is a new concept generated, add it to the PMSD architecture and set it as a private concept.

For each concept, declare or relate his or her own knowledge, document, and data about that concept and put them as the elements of the content associated with that concept.

Therefore, a PMSD is the picture that depicts a snapshot of a stakeholder's perception of concepts. It embodies his or her related purposes, context, and content. In a collaboration-support system, a PMSD is represented as XML (extensible markup language) formats to facilitate analysis.

The third step is to conduct the perspective analysis. By comparing and analyzing stakeholders' perspective models, it is possible to determine the degree of agreement among their opinions during their interaction. As shown in Figure 4, given the PMSDs for certain stakeholders, we can ask them to review others' perspective models. The review information is used to compare the perspective models and determine the similarity of two stakeholders' perspectives toward a shared concept. We can also aggregate multiple stakeholders' perspective models and compare their general attitudes at different levels of abstraction. Furthermore, we can track the evolution of the perspective model based on the clustering analysis results. The procedure is called perspective analysis (Figure 4).

The first step is to determine the inconsistency (i.e., the distance) among a group of perspective models. There are two approaches: the intuitive approach and the analytical approach. The intuitive approach relies on the insights of the stakeholders. The analytical approach uses mathematical algorithms to derive the distance through positional analysis, which is based on a formal method used in social network analysis (Wasserman & Faust, 1994). This approach views the perspective models of a group of stakeholders toward a single concept as a network of opinions associated with each other. In this network, a stakeholder, who possesses a perspective model, has relationships with others' perspective models. We define these relationships as their perceptional attitudes toward each other. A group of perspective models toward a given concept are placed as a graph (i.e., a PM network). Two perspective models are compatible (or similar) if they are in the same position in the network structure. In social network analysis, position refers to a collection of individuals who are similarly embedded in networks of relations. If two perspective models are structurally equivalent (i.e., their relationships with other perspective models are the same), we assume that they are purely compatible and there are no detectable differences. That implies that they have the same perception toward others, and others have same perception toward them.

A distance matrix is derived for each PM network. It represents the situation of perspective compatibility among a group of stakeholders for a given concept. We can also compare stakeholders' perspective models for multiple concepts by measuring the structural equivalence across the collection of perspective model networks. Perspective distance matrices serve as the basis for cluster analysis. Hierarchical clustering is a data analysis technique that is suited for partitioning the perspective models into subclasses. It groups entities into subsets so that entities within a subset are relatively similar to each other. Hierarchical clustering generates a tree structure (or a dendrogram), which shows the grouping of the perspective models. It illustrates that the perspective models are grouped together at different levels of abstraction (Figure 4).

The cluster tree exposes interesting characteristics of the social interactions. Within a collaborative project, the participants of the organization cooperate and build the shared reality (i.e., the common understanding of the stakeholders toward certain concepts) in the social interaction process (Berger & Luckman, 1966). Understanding the process of building shared realities is the key to managing social interactions. The shared reality can be represented by the abstraction of close perspective models among a group of stakeholders. As a matter of fact, the cluster tree depicts the structures of the shared reality since a branch of the clustering tree at a certain level implies an abstract perspective model with certain granularity. The height of the branch indicates the compatibility of the leaf perspective models. A cluster tree with simple structure and fewer levels implies that all of the perspective models have similar attitudes (or positions) toward others.

Figure 4. The perspective analysis procedure

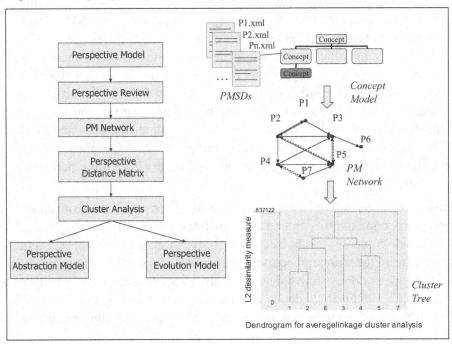

While the perspective models are changing, the clustering analysis can be used as a systematic way to depict the transformation of the perspective models. The change of the cluster trees at different stages of collaboration reveals the characteristics of perspective evolution. Investigating the changes of the topological patterns of the clustering trees leads to ways to interfere in the perspective evolutions.

Conflict Management

Given the condition that the social interactions are analytically measured, control mechanisms can be derived to manage the evolutions of the perspective models and therefore to support collaboration. Theses mechanisms could be selected and used by the group managers or coordinators to control conflicts. They can be classified into the following strategies.

Process Control

The perspective analysis can be performed for all of the stakeholders who might act on or influence a task. By evaluating their perspective compatibility and the execution feasibility of future tasks, which are in the plan but have not been conducted yet, we can prevent some conflicts by noticing their potential existence earlier. By providing certain information to stakeholders, it is possible to change the perception matrix and therefore to increase the perspective consistency of a task. It is possible to directly adjust the sequences and dependencies among the tasks to maintain the integrity of the opinions of stakeholders.

Perspective Control and Ontology Control

First, it is possible to directly influence stakeholders' perspectives (their content, purpose, and context) to maintain the integrity and compatibility of the opinions toward a certain concept or task. Analyzing social interactions will identify the perspective models with low similarities and reveal the conflicts clearly. Thus, we can focus on the stakeholders who have singular perspectives and understand their rationale. Second, communication channels can be built to increase the interaction opportunities among stakeholders with different perspective models. The group can manipulate the concept structure through clarifying the meanings and definitions of critical concepts so that people have shared understanding. It is also feasible to serve stakeholders with different concepts to isolate their perspectives. An opposite way is to use conflicting perspectives as means to enhancing brainstorming and innovation. Third, strategies can be derived to manage the conflicts through influencing stakeholders' information access and comprehension. Possible solutions include providing suitable trainings based on their perspectives and the job requirements, assisting the critical stakeholder to review the relevant information during certain conflicting tasks, and recording the discussions about the shared concept for future reuse.

Organization Control

The clustering tree shows the grouping features of stakeholders' perspectives. Using different organizational structures will change the communication channels and the perception distances. If two stakeholders are separated into different groups, the possibility of interaction will decrease. We can change the task assignment or modify stakeholder' roles to affect their contexts. It is even possible to add or remove stakeholders associated with a certain task to avoid the conflicting situation or to move the stakeholders with similar perspectives together.

Data and Information Control

This control mechanism is to affect the conflicts through appropriately providing and handling external data and information that will be accessed by the stakeholders. Examples are to use consistent checking and version-control mechanisms to maintain the product data integrity, to track the changes of shared data and information by referencing to the perspective changing, and to map the shared data and information to perspective models so that the system realizes the specific impact of the conflicts toward the working results.

Building Electronic Collaboration Support Systems Using the Perspective Modeling Approach

The perspective modeling and analyzing methodology provides a theoretical basis for building new knowledge management systems. The STARS system is a prototype system to support collaboration over the Internet. It is also developed as an experimental apparatus for testing the research. The system implements the process modeling, perspective modeling, and sociotechnical analysis methodologies. On the other hand, it collects process and perspective data once stakeholders use it as a collaboration tool. By investigating the collected experimental data, we can determine the effectiveness of the approach and therefore improve it.

The STARS system provides a Web-based environment that supports the collaboration process representation, conflict management, and knowledge integration within a project team. Stakeholders declare, share, and modify their perspective models on the Web. The perspectives models are analyzed in the system and stakeholders' roles in the collaboration tasks are depicted.

Figure 5. STARS system architecture

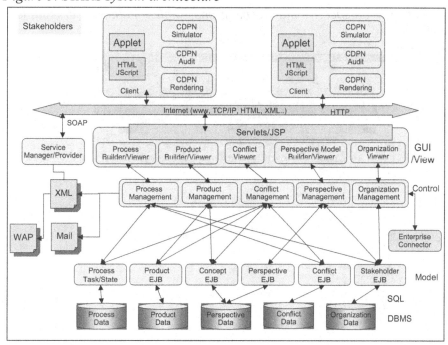

The system implements the functional modules (e.g., perspective management, process management, conflict management, etc.) by using J2EE1.4 and Web services technologies (Figure 5). It provides methods to detect, analyze, and track the conflicts during collaboration. It also supports the business-to-business process communications through SOAP and UDDI.

Figure 6 shows the knowledge perspective management module that allows stakeholders to declare and review their perspective information according to a concept structure tree. The system can analyze the perspective models, detect and predict conflicts, and suggest possible control strategies. The process management system of STARS uses an XML-based process modeling tool for process planning, scheduling, simulation, and execution. It helps the stakeholders notice what is happening and who is doing what at any time. Stakeholders declare their perspectives during each step of the process. The system determines the conflict ratio of each task based on the perspective analysis.

Groups of designers, business analysts, and consultants working in a U.S. national construction research institute have been using STARS in their

Figure 6. The perspective-management and conflict-management modules of STARS

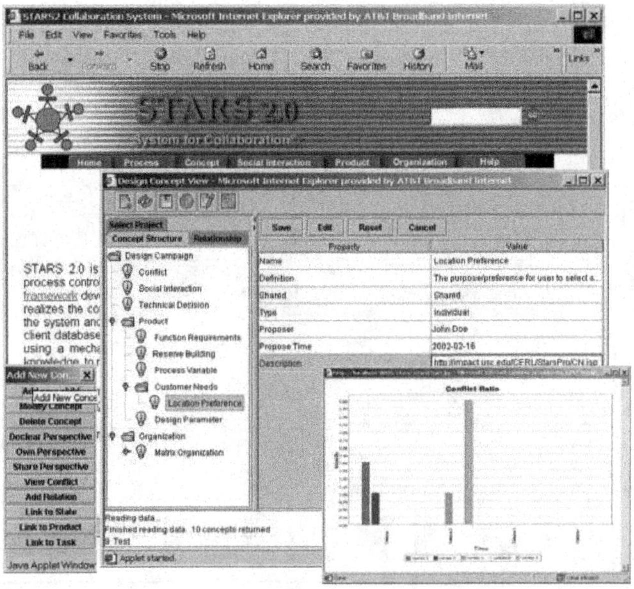

Figure 7. An example of detecting conflicts from perspective analysis

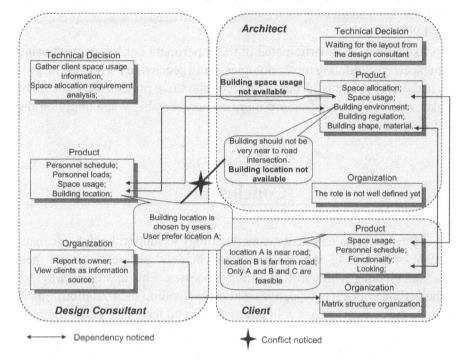

small projects. Feasibility and computability of the analysis algorithms were proved. Figure 7 depicts an example of using STARS to solve a conflict problem through perspective analysis. Before using STARS, similar cases as described below often happened in one design team:

Within a design project, at the first meeting, the client's design consultant stated that the building was to be placed at a location on the site. The architect listened to the client's reasoning but noted that this location is not ideal from either an aesthetic or a functional point of view, since it would be too close to a major road intersection.

The STARS perspective analyzing functions helped users notice the dependencies and differences of views among the stakeholders. The conflict was detected by tracking and mapping the perspective models of the three stakeholders. STARS compared the perspective models at an early stage of the design. Although there was no direct meeting between the design consultant and the architect, the system detected a potential conflict during the design process.

The stakeholders who participated in the experiment considered that using the perspective modeling methodologies could accelerate their learning process and detect conflicts earlier in their collaborative projects. The causes of breakdowns of collaboration are more comprehensible when applying the analysis methodologies.

Conclusion

This chapter presents a systematic methodology to support knowledge management by modeling and analyzing stakeholders' perspectives and their social interactions within collaborative processes. This approach provides methods for capturing perspectives and understanding their relationships to facilitate the control of the evolution of the shared insights. It avails knowledge management and conflict management by systematically facilitating the manipulation of the process, the perspectives, the organizational structure, and the shard data and information. The STARS system was built to improve the coordination among stakeholders. Its perspective modeling function provides an efficient way for stakeholders to understand the meanings and improve coordination during their collaboration over the Internet.

This research has some limitations. First, the closed-loop perspective management methodology requires stakeholders to be actively involved in the building and updating of perspective models. This might be overkill when the group is already very efficient and stable. Second, using the perspective analysis requires the computing tool and thus introduces a higher level of complexity. The system users have to be able to honestly and clearly specify their understandings toward the concepts and others' perspectives. In the future, the perspective analysis model can be improved by applying advanced statistics and econometrics techniques. It is also important to generate dynamic modeling methods to define the relationships between the evolution of perspective models and the quality of online collaboration.

References

Alavi, M., & Leidner, D. E. (2001). Review: Knowledge management and knowledge management systems: Conceptual foundations and research issues. *MIS Quarterly, 25*(1), 105-136.

Arias, E. G., Eden, H., Fischer, G., Gorman, A., & Scharff, E. (2000). Transcending the individual human mind-creating shared understanding through collaborative design. *ACM Transactions on Computer-Human Interaction, 7*(1), 84-113.

Becerra-Fernanaez, I., & Sabherwal, R. (2001). Organizational knowledge management: A contingency perspective. *Journal of Management Information Systems, 18*(1), 23-55.

Berger, P., & Luckman, T. (1966). *The social construction of reality a treatise in the sociology of knowledge.* New York: Doubleday.

Briggs, R. O., Vreede, G.-J., & Nunamaker, J. F., Jr. (2003). Collaboration engineering with thinkLets to pursue sustained success with group support systems. *Journal of Management Information Systems, 19*(1), 31-64.

Carley, K. M., & Prietula, M. J. (1994). ACTS theory: Extending the model of bounded rationality. In *Computational organization theory* (pp. 55-88). UK: Lawrence Erlbaum Associates.

Chae, B., Koch, H., Paradice, D., & Huy, V. V. (2005). Exploring knowledge management using network theories: Questions, paradoxes, and prospects. The Journal of Computer Information Systems, 45(4), 62-15.

Chung, C.-W., Kim, C.-R., & Dao, S. (1999). Knowledge and object-oriented approach for interoperability of heterogeneous information management systems. *Journal of Database Management, 10*(3), 13-25.

Clancey, W. J. (1993). Guidon-manage revisited: A socio-technical systems approach. *Journal of Artificial Intelligence in Education, 4*(1), 5-34.

Clancey, W. J. (1997). The conceptual nature of knowledge, situations, and activity. In P. Feltovich, R. Hoffman, & K. Ford (Eds.), *Human and machine expertise in context* (pp. 247-291). CA: AAAI Press.

Dym, C. L., & Levitt, R. E. (1991). Toward the integration of knowledge for engineering modeling and computation. *Engineering with Computers, 7*(1), 209-224.

Earl, M. J. (2001). Knowledge management strategies: Toward a taxonomy. *Journal of Management Information Systems, 18*(1), 215-233.

Easley, R. F., Sarv, D., & Crant, J. M. (2003). Relating collaborative technology use to teamwork quality and performance: An empirical analysis. *Journal of Management Information Systems, 19*(4), 247-268.

Erickson, T., & Kellogg, W. A. (2000). Social translucence: An approach to designing systems that support social processes. *ACM Transactions on Computer-Human Interactions, 7*(1), 59-83.

Hanson, M., Nohira, N., & Tierney, T. (1999). What is your strategy for managing knowledge? *Harvard Business Review*, 106-116.

Hardjono, T. W., & van Marrewijk, M. (2001). The social dimensions of business excellence. *Corporate Environmental Strategy, 8*(3), 223-233.

Huhns, M. N., & Stephens, L. M. (1999). Personal ontologies. *IEEE Internet Computing, 3*(5), 85-87.

Kannapan, S., & Taylor, D. (1994). The interplay of context, process, and conflict in concurrent engineering, *Journal of Concurrent Engineering Research and Applications, 2*(1), 183-196.

Kwan, M. M., & Balasubramanian, P. (2003). Process-oriented knowledge management: A case study. *Journal of Operational Research Society, 54*(1), 204-211.

Lee, H., & Choi, B. (2003). Knowledge management enablers, processes, and organizational performance: An integrative view and empirical examination. *Journal of Management Information Systems, 20*(1), 179-228.

Lu, S. C.-Y., & Cai, J. (2001). A collaborative design process model in the sociotechnical engineering design framework. *Artificial Intelligence for Engineering Design, Analysis and Manufacturing, 15*(1), 3-20.

Nonaka, I., Reinmoeller, P., & Senoo, D. (1998). The "ART" of knowledge: Systems to capitalize on market knowledge. *European Management Journal, 16*(6), 673-684.

Nonaka, I., & Takeuchi, H. (1995). *The knowledge-creating company.* New York: Oxford University Press.

O'Leary, D. E. (1998). Enterprise knowledge management. *IEEE Computer*, 54-61.

Preece, A., Flett, A., & Sleeman, D. (2001). Better knowledge management through knowledge engineering. *IEEE Intelligent Systems*, 36-43.

Reimer, U., Margelisch, A., & Staudt, M. (2000). EULE: A knowledge-based system to support business processes. *Knowledge-Based Systems, 13*, 261-269.

Rouse, W. B. (2001). Need to know: Information, knowledge, and decision making. *IEEE Transactions on Systems, Man, and Cybernetics. Part C: Applications and Reviews, 32*(4), 282-292.

Shaw, D., Ackermann, F., & Eden, C. (2003). Approaches to sharing knowledge in group problem structuring. *Journal of the Operational Research Society, 54*, 936-948.

Siau, K. (1999). Information modeling and method engineering: A psychological perspective. *Journal of Database Management, 10*(4), 44-50.

Sowa, J. F., & Zachman, J. A. (1992). Extending and formalizing the framework for information systems architecture. *IBM System Journal, 31*(3), 590-616.

Spender, J. C. (1996). Making knowledge the basis of a dynamic theory of the firm. *Strategic Management Journal, 17*, 45-62.

Staab, S., Schnurr, H.-P., Studer, R., & Sure, Y. (2001). Knowledge processes and ontologies. *IEEE Intelligent Systems*, 26-34.

Tanriverdi, H. (2005). Information technology relatedness, knowledge management capability, and performance of multibusiness firms. *MIS Quarterly, 29*(2), 311-335.

Vet, P. E., & Mars, N. J. (1998). Bottom-up construction of ontologies. *IEEE Transaction on Knowledge and Data Engineering, 10*(4), 513-526.

Wasserman, S., & Faust, K. (1994). *Social network analysis: Methods and applications.* New York: Cambridge University Press.

Zack, M. H. (1999). Managing codified knowledge. *Sloan Management Review, 40*(4), 45-58.

Chapter VIII

Modality of Business Rules

Terry Halpin, Neumont University, USA

Abstract

A business domain is typically subject to various business rules. In practice, these rules may be of different modalities (e.g., alethic and deontic). Alethic rules impose necessities, which cannot, even in principle, be violated by the business. Deontic rules impose obligations, which may be violated, even though they ought not to be. Conceptual modeling approaches typically confine their specification of constraints to alethic rules. This chapter discusses one way to model deontic rules, especially those of a static nature. A formalization based on modal operators is provided, and some challenging semantic issues are examined from both logical and pragmatic perspectives. Because of its richer semantics, the main graphic notation used is that of object-role modeling (ORM). However, the main ideas could be adapted for UML and ER as well. A basic implementation of the proposed approach has been prototyped in Neumont ORM Architect (NORMA), a software tool that supports automated verbalization of both alethic and deontic rules.

Introduction

In the wider sense, an information system corresponds to a business domain or universe of discourse rather than an automated system. Business domains are constrained by various business rules, which specify required or desirable states of affairs or behavior. Business rules may be of different modalities (e.g. alethic and deontic). Alethic rules impose necessities, which cannot, even in principle, be violated by the business, typically because of some physical or logical law. For example, each employee was born on at most one date, or no product is a component of itself. Deontic rules impose obligations, which may be violated, even though they ought not to be. For example, it is obligatory that each employee is married to at most one person, and it is forbidden that any person smokes in any office.

Various information modeling approaches exist for modeling business domains at a high level, for example, entity-relationship (ER) modeling (Chen, 1976), the unified modeling language (UML; Object Management Group [OMG], 2003a, 2003b; Rumbaugh, Jacobson, & Booch, 1999), and object-role modeling (ORM; Halpin, 1989, 2001, 2006). However, these modeling approaches typically confine their specification of rules to those of an alethic modality, ignoring deontic rules. A notable exception is the proposal of Krogstie and Sindre (1996) to extend the Tempora approach to capture not only alethic rules (necessities) and deontic rules (obligations), but also recommendations (in their proposal, they include recommendations as a subclass of deontic rules, but we classify recommendations in terms of a different and weaker modality that is not discussed further here). While our approach is similar to that of Krogstie and Sindre in drawing upon the formalism of deontic logic, it covers new ground by considering the automated verbalization of deontic rules, applying the ideas within the context of ORM, and examining embedded deontics and other logical issues.

It is important for a business to have a clear understanding of all its rules, including deontic ones, whether or not the business chooses to enforce these rules or monitor violations of them by means of an automated system. In recognition of this need, as well as to facilitate the exchange of semantics between businesses, the OMG is currently finalizing a proposal to specify a business semantics layer on top of its software-specific layers (OMG, 2006).

The proposal that was accepted by the OMG for finalization is the Semantics of Business Vocabulary and Rules (SBVR) submission. As a contributor to this submission, the author focused on the formal logic underpinnings of

SBVR. This chapter relates in part to that fragment of his contribution that is concerned with the modeling of deontic rules, especially those of a static nature. Because of its richer semantics, the main graphic notation used is that of ORM 2 (the next generation of object-role modeling). However, the main ideas could be adapted for UML and ER as well.

The next section provides a simple overview of the use of modal operators in expressing business rules of alethic and deontic modalities, and illustrates the automated verbalization of these rules as implemented in a prototype ORM 2. The section after that discuses the formal underpinnings of static, alethic rules. The following section does likewise for static, deontic rules, and examines some challenging semantic issues from both logical and pragmatic perspectives. The subsequent section briefly raises some issues relating to dynamic rules. The final section summarizes the main results, suggests topics for future research, and lists references.

Modal Operators and Rule Verbalization

Business constraint formulations may use any of the basic alethic or deontic modal operators from modal logic, as shown in Table 1. These modal operators are treated as proposition-forming operators on propositions (rather than actions). Other equivalent readings may be used in whatever concrete syntax is used to originally declare the rule (e.g., necessary might be replaced by required, and obligatory might be replaced by "ought to be the case"). Derived modal operators may also be used in the surface syntax, but are translated into the basic modal operators plus negation (~). For example, "it is impossible that p" is defined as "it is not possible that p" ($\sim \Diamond p$), and "it is forbidden that p" is defined as "it is not permitted that p" ($Fp =_{df} \sim Pp$).

Table 1. Alethic and deontic modal operators

Alethic		Deontic	
Reading	*Symbol*	*Reading*	*Symbol*
It is necessary that	□	It is obligatory that	*O*
It is possible that	◊	It is permitted that	*P*

The following modal negation rules apply: it is not necessary that ≡ it is possible that not ($\sim\Box p \equiv \Diamond\sim p$); it is not possible that ≡ it is necessary that not ($\sim\Diamond p \equiv \Box\sim p$); it is not obligatory that ≡ it is permitted that it is not the case that ($\sim\boldsymbol{O}p \equiv \boldsymbol{P}\sim p$); it is not permitted that ≡ it is obligatory that it is not the case that ($\sim\boldsymbol{P}p \equiv \boldsymbol{O}\sim p$). In principle, these rules could be used with double negation to get by with just one alethic modal operator (e.g., $\Diamond p$ could be defined as $\sim\Box\sim p$, and $\boldsymbol{P}p$ could be defined as $\sim\boldsymbol{O}\sim p$).

ORM is a conceptual modeling approach that models any business domain in terms of objects (entities or values) that play roles in relationships (unary, binary, or longer), also known as facts, relegating the attribute construct merely to derived views, and hence offering greater semantic stability than attribute-based approaches like ER and UML (DeTroyer & Meersman, 1995; Halpin, 2001, 2004; ter Hofstede, Proper, & Weide, 1993). ORM also has a rich graphic notation for capturing constraints, which ORM tools can transform into implementation code for enforcement. In ORM 2 (Halpin, 2005c), the latest version of ORM, each constraint or rule has an associated modality, determined by the logical modal operator that functions explicitly or implicitly as its main operator. ORM 2 distinguishes between positive, negative, and default verbalizations of rules (Halpin, 2004). In positive verbalizations, an alethic modality of necessity is often assumed (if no modality is explicitly specified), but may be explicitly proposed. For example, the following static constraint:

C1   **Each** Person was born in **at most one** Country.

may be explicitly verbalized with an alethic modality, thus

C1'   **It is necessary that each** Person was born in **at most one** Country.

We interpret this in terms of possible world semantics, as introduced by Saul Kripke and other logicians in the 1950s. A proposition is necessarily true if and only if it is true in all possible worlds. With respect to a static constraint declared for a given business domain, a possible world corresponds to a state of the fact model that might exist at some point in time. The constraint C1 means that for each state of the fact model, each instance in the population of Person is born in at most one country.

A proposition is possible if and only if it is true in at least one possible world. A proposition is impossible if and only if it is true in no possible world (i.e., it is false in all possible worlds). In ORM, constraint C1 may be reformulated as the following negative verbalization:

C1" **It is impossible that the same** Person was born in **more than one** Country.

In practice, both positive and negative verbalizations are useful for validating constraints with domain experts, especially when illustrated with sample populations that provide satisfying examples or counterexamples respectively. For example, Figure 1 models a birth association in (a) ORM, (b) the popular Barker (1990) version of ER, and (c) UML. In ORM, object types (e.g., Person, Country) are depicted as named, soft rectangles (earlier versions of ORM used ellipses instead). A logical predicate is depicted as a named sequence of role boxes, each of which is connected by a line segment to the object type whose instances may play that role. The combination of a predicate and its object types is a fact type, which is the only data structure in ORM.

From an ORM perspective, the left role of the binary fact type "Person was born in Country" has two alethic constraints applied. The bar over the role depicts an alethic uniqueness constraint verbalized in positive form as "Each Person was born in at most one Country." The satisfying fact population shown in the fact table immediately below the fact type illustrates this constraint (the person entries are unique) as well as the lack of a uniqueness constraint on the right-hand role (a country entry is duplicated). The counterpopulation in the fact table illustrates how to violate the uniqueness constraint by providing a counterworld where it is possible for the same person to be born

Figure 1. A birth association modeled in (a) ORM, (b) Barker ER, and (c) UML

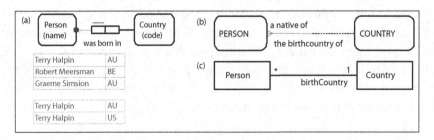

in more than one country (directly contradicting the negative verbalization of the constraint). The solid dot in Figure 1a depicts the alethic mandatory role constraint that may be verbalized as "Each Person was born in some Country."

In Barker ER, the presence and absence of a uniqueness constraint is depicted by using the crow's-foot notation (for many), and the mandatory constraint is depicted as a solid rather than dashed line. In UML the constraints are captured as multiplicity constraints where "*" denotes zero or more. One advantage of the ORM constraint notation is that it extends readily to associations of higher arity (e.g., ternary or quaternary associations), whereas the Barker notation does not extend at all and the UML notation breaks down in many cases (Halpin, 2004b).

As an example of a ternary association that can be handled in UML, consider the room-booking example in Figure 2. In ORM, a uniqueness constraint over multiple roles applies to the combination of those roles. For example, the alethic uniqueness constraint over the first two roles of the fact type "Room at HourSlot was booked for Course" may be verbalized in positive form as "For each Room and HourSlot, that Room at that HourSlot is booked for more than one Course," and in negative form as "It is impossible that the same Room at the same HourSlot is booked for more than one Course." The fact table illustrates this constraint with a satisfying fact population.

Many business constraints are deontic rather than alethic in nature. To avoid confusion, when declaring a deontic constraint, the deontic modality should always be explicitly included. Consider the following static, deontic constraint.

Figure 2. A ternary association in (a) ORM and (b) UML

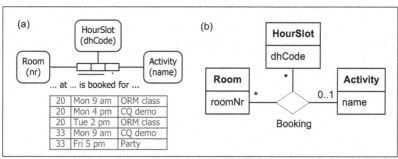

C2   It is obligatory that each Person is a husband of **at most one** Person.

If this rule were instead expressed simply as "Each Person is a husband of at most one Person," it would not be obvious that a deontic interpretation was intended. The deontic version indicates a condition that ought to be satisfied while recognizing that the condition might not be satisfied. Including the obligation operator makes the rule much weaker than a necessity claim since it allows that there could be some states of the fact model where a person is a husband of more than one wife (excluding same-sex unions from instances of the husband relationship). For such cases of polygamy, it is important to know the facts indicating that the person has multiple wives. Rather than reject this possibility, we allow it and then typically perform an action that is designed to minimize the chance of such a situation arising again (e.g., send a message to inform legal authorities about the situation). In ORM, constraint C2 may be reformulated as the following negative verbalization:

Figure 3. Screenshot from NORMA, showing positive verbalization of some constraints

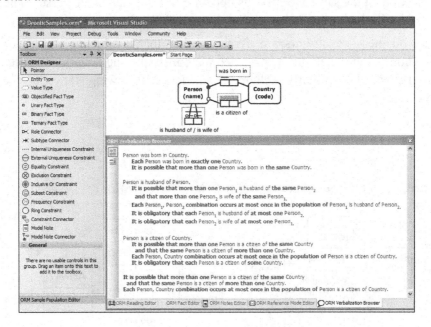

C2'   **It is forbidden that the same** Person is a husband of **more than one** Person.

Figure 3 shows a screenshot from NORMA (Neumont ORM Architect), illustrating positive verbalization of some alethic and deontic constraints in ORM 2. Alethic constraints are colored violet, while deontic constraints are colored blue. In addition, deontic constraints are distinguished by a small *o* (for *obligatory*). The citizenship and marriage fact types have spanning uniqueness constraints, and hence are alethically many-to-many associations. However, each role of the marriage association has a deontic uniqueness constraint (e.g., "It is obligatory that each Person$_1$ is husband of at most one Person$_2$"). Subscripts may be used to distinguish object variables of the same type. If the mandatory-role dot is open rather than solid, the mandatory constraint is deontic (e.g., "It is obligatory that each Person is a citizen of some Country").

Figure 4 displays another screenshot from NORMA, illustrating negative verbalization of a deontic uniqueness constraint spanning the first two roles of the ternary fact type "Room at HourSlot is booked for Activity." The constraint verbalization ("It is forbidden that the same Room at the same HourSlot is booked for more than one Activity") uses the deontic $F(\sim P)$ operator. All verbalizations in NORMA are performed automatically via XSLT transforms, and hence may be readily adapted for different native languages. NORMA itself is an open-source plug-in to Visual Studio .NET 2005, and may be downloaded from http://sourceforge.net/projects/orm.

In practice, most business rules include only one modal operator, and this operator is the main operator of the whole rule expression. For these cases,

Figure 4. NORMA screen shot illustrating negative verbalization of a deontic constraint

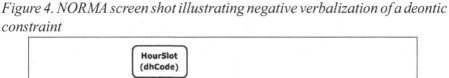

we simply tag the constraint as being of the modality corresponding to its main operator, without committing to any particular modal logic. Apart from this modality tag, there are some basic modal properties that may be used in transforming the original high-level expression of the rule into a standard logical formulation. At a minimum, these include the modal negation rules.

We also make use of equivalences that allow one to move the modal operator to the front of the formula. For example, suppose the user formulates rule C1 instead as:

For each Person, **it is necessary that that** Person was born in **at most one** Country.

The modal operator is now embedded in the scope of a universal quantifier. To transform this rule to a standard logical formulation that classifies the rule as an alethic necessity, we move the modal operator before the universal quantifier, to give:

It is necessary that each Person was born in **at most one** Country.

For such tasks, we assume that the Barcan formulae and their converses apply, so that □ and ∀ are commutative, as are ◊ and ∃. In other words,

$$\forall x \Box Fx \equiv \Box \forall x Fx$$
$$\exists x \Diamond Fx \equiv \Diamond \exists x Fx.$$

While these commutativity results are valid for all normal, alethic modal logics, some philosophical concerns have been raised about these equivalences, for example, see Sections 4.6 to 4.8 of Girle (2000).

As a deontic example, suppose the user formulates rule C2 instead as:

For each Person, **it is obligatory that that** Person is a husband of **at most one** Person.

Using a deontic variant of the Barcan equivalences, we commute the ∀ and *O* operators, thus transforming the rule to the deontic obligation:

It is obligatory that each Person is a husband of **at most one** Person.

So far, our rule examples have included just one modal operator, which (perhaps after transformation) also turns out to be the main operator. Ignoring dynamic aspects, we may handle such cases without needing to commit to the formal semantics of any specific modal logic. The only impact of tagging a rule as a necessity or obligation is on the rule enforcement policy. Enforcement of a necessity rule should never allow the rule to be violated. Enforcement of an obligation rule should allow states that do not satisfy the rule condition, and take some other remedial action. The precise action to be taken is not specified here, but the tool's default is to generate a message when an update violates the rule.

At any rate, a business person ought to be able to specify a deontic rule first at a high level, without committing at that time to the precise action to be taken if the condition is not satisfied; of course, the action still needs to be specified later in refining the rule to make it fully operational.

Static, Alethic Constraints

Rule formulations may make use of two alethic modal operators: □ = it is necessary that, and ◊ = it is possible that. Static constraints are treated as alethic necessities by default, where each state of the fact model corresponds to a possible world. Given the fact type "Person was born in Country," the constraint "Each Person was born in at most one Country" is equivalent to the logical formulation $\forall x$:Person $\exists^{0..1} y$:Country x was born in y. This formula is understood to be true for each state of the knowledge base. Pragmatically, the rule is understood to apply to all future states of the fact model until the rule is revoked or changed. This understanding could be made explicit by proposing the formula with □ to yield the modal formula $\Box \forall x$:Person $\exists^{0..1} y$: Country x was born in y. For compliance with common logic (ISO, 2005), such formulae could then be treated as irregular expressions, with the modal necessity operator treated as an uninterpreted symbol (e.g., using [N] for □).

However, we leave this understanding as implicit and do not commit to any particular modal logic.

For the model theory, we omit the necessity operator from the formula. Instead, we merely tag the rule as a necessity. The implementation impact of the alethic necessity tag is that any attempted change that would cause the model of the business domain to violate the constraint must be dealt with in a way that ensures the constraint is still satisfied (e.g., reject the change, or take some compensatory action).

Typically, the only alethic modal operator in an explicit rule formulation is $\Box$, and this is at the front of the rule. This common case was covered earlier. If an alethic modal operator is placed elsewhere in the rule, we first try to normalize it by moving the modal operator to the front, using transformation rules such as the modal negation rules ($\sim\Box p \equiv \Diamond\sim p$; $\sim\Diamond p \equiv \Box\sim p$) and/or the Barcan formulae and their converses ($\forall x\Box\Phi x \equiv \Box\forall x\Phi x$ and $\exists x\Diamond\Phi x \equiv \Diamond\exists x\Phi x$, i.e., $\Box$ and $\forall$ are commutative, as are $\Diamond$ and $\exists$). For example, the embedded formulation $\forall x$:Person $\Box\exists^{0..1}y$:Country x was born in y (For each Person, it is necessary that that Person was born in at most one Country) may be transformed into $\Box\forall x$:Person $\exists^{0..1}y$:Country x was born in y (It is necessary that each Person was born in at most one Country).

We also allow use of the following equivalences: $\Box\Box p \equiv \Box p$, $\Diamond\Diamond p \equiv \Diamond p$, $\Box\Diamond\Box\Diamond p \equiv \Box\Diamond p$, and $\Diamond\Box\Diamond\Box p \equiv \Diamond\Box p$. These hold in S4, but not in some modal logics, for example, K or T (Girle, 2000).

Though not supported by NORMA, the SBVR proposal also allows a single rule to include multiple occurrences of modal operators, including the nesting of a modal operator within the scope of another modal operator. While this expressiveness may be needed to capture some rare but real business rules, it complicates attempts to provide a formal semantics.

In extremely rare cases, a formula for a static business rule might contain an embedded alethic modality that cannot be eliminated by transformation. For such cases, we could retain the modal operator in the rule formulation and adopt the formal semantics of a particular modal logic. There are many normal modal logics to choose from (e.g., K, K4, KB, K5, DT, DB, D4, D5, T, Br, S4, S5) as well as many abnormal modal logics (e.g., C2, ED2, E2, S0.5, S2, S3). For a discussion of these logics and their interrelationships, see Girle (2000, pp. 48, 82). For SBVR, if we decide to retain the embedded alethic operator for such cases, we choose S4 for the formal semantics. The possibility of schema evolution along with changes to necessity constraints may seem to violate S4, where the accessibility relationship between possible worlds is transitive, but we resolve this by treating such evolution as a metalevel concern. Alternatively, we may handle such very rare cases by moving the embedded alethic operators down to domain-level predicates (e.g., "is necessary") in a similar fashion to the way we deal with embedded deontics (see later).

Static, Deontic Rules

Constraint formulations may make use of the standard deontic modal operators (O = it is obligatory that; P = it is permitted that) as well as F = it is forbidden that (defined as $\sim P$, i.e., "it is not permitted that"). If the rule includes exactly one deontic operator, O, and this is at the front, then the rule may be formalized as Op, where p is a first-order formula that is tagged as obligatory (rather than necessary). For NORMA, this tag is assigned only the following informal semantics: it ought to be the case that p (for all future states of the fact model, until the constraint is revoked or changed). The implementation impact is that it is possible to have a state in which the rule's condition is violated (i.e., not satisfied), in which case some appropriate action (e.g., messaging) ought to be taken to help reduce the chance of future violations. Later work will address rule enforcement, including the specification of appropriate actions in response to deontic rule violations.

From a model-theoretic perspective, a model is an interpretation where each nondeontic formula evaluates to true, and the model is classified as a permitted model if the p in each deontic formula (of the form Op) evaluates to true; otherwise, the model is a forbidden model (though it is still a model). Note that this approach removes any need to assign a truth value to expressions of the form Op.

Figure 5.Deontic constraints obligate the marriage relationship to be 1:1

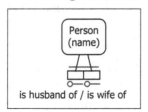

Recall our current marriage example where the fact type "Person is husband of Person" is declared to be many to many, but each role of this fact type has a deontic uniqueness constraint to indicate that the fact type ought to be 1:1 (see Figure 5). The deontic constraint on the husband role verbalizes as "It is obligatory that each Person is husband of at most one Person." This formalizes as $O\forall x$:Person $\exists^{0..1}y$:Person x is husband of y, which may be captured by entering the rule body as $\forall x$:Person $\exists^{0..1}y$:Person x is husband of y, and tagging the rule as deontic. The other deontic constraint (each wife should have at most one husband) may be handled in a similar way.

In this example, the combination of alethic and deontic constraints is consistent, but this is not always the case. For example, the argument (role set) of a deontic uniqueness constraint must be a proper subset of the argument of an alethic uniqueness constraint. For instance, if the marriage predicate is alethically 1:1, then no deontic uniqueness constraint may be added (if something is already necessary, it makes no sense to declare it obligatory).

Some formulae allowed by SBVR are illegal in some deontic logics (e.g., iterating modal operators such as OPp is forbidden in von Wright's deontic logic), and deontic logic itself is "rife with disagreements about what should be the case" (Girle, 2000, p. 173).

If a deontic modal operator is embedded later in the rule formulation, we first try to normalize the formula by moving the modal operator to the front, using transformation rules such as $p \supset Oq .\equiv. O(p \supset q)$ or deontic counterparts to the Barcan formulae.

In some cases, a formula for a static business rule might contain an embedded deontic modality that cannot be eliminated by transformation. In this case, we still allow the business user to express the rule at a high level using such embedded deontic operators, but where possible we transform the formula to a first-order formula without modalities by replacing the modal operators with predicates at the business domain level. These predicates (e.g., "is forbidden") are treated like any other predicate in the domain except that their names are reserved, and they are given some basic additional formal semantics to capture the deontic modal negation rules: it is not obligatory that ≡ it is permitted that it is not the case that ($\sim Op \equiv P\sim p$), and it is not permitted that ≡ it is obligatory that it is not the case that ($\sim Pp \equiv O\sim p$). For example, these rules entail an exclusion constraint between the predicates "is forbidden" and "is permitted."

This latter approach may also be used as an alternative to tagging a rule as deontic, thereby (where possible) moving deontic aspects out of the metamodel and into the business domain model. For example, consider the following rule:

Car rentals ought not be issued to people who are barred drivers at the time the rental was issued.

This deontic constraint may be captured by the following ORM textual constraint on the domain fact type "CarRental is forbidden":

CarRental is forbidden **if**
>   CarRental was issued at Time **and**
>   CarRental was issued to Person **and**
>   Person is a barred driver at Time.

The fact type "Person is a barred driver at Time" is derived from other base fact types (Person was barred at Time, Person was unbarred at Time) using the ORM derivation rule:

Figure 6.Forbidding rentals to barred drivers using a domain-level predicate

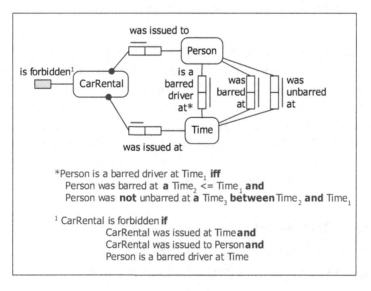

Person is a barred driver at Time1 **iff**

  Person was barred at **a** Time$_2$ <= Time$_1$ **and**

  Person was **not** unbarred at **a** Time$_3$ **between** Time$_2$ **and** Time$_1$.

The deontic constraint may be formalized by the first-order formula $\forall x$: CarRental $\forall y$:Person $\forall t$:Time [(x was issued at t & x was issued to y & y is a barred driver at t) $\supset x$ is forbidden]. This schema (see Figure 6) allows for the possible existence of forbidden car rentals; if desired, some fact types could be added to describe actions (e.g., sending messages) to be taken in reaction to such an event.

For other examples illustrating this approach, including the use of derivation rules and objectification, see the SBVR submission to the OMG. Our approach to objectification works for those cases where a fact (proposition taken to be true) is being objectified (which covers the usual cases of nominalization; Halpin, 2005b), but it does not handle cases where no factual claim is being made of the proposition.

SBVR is intended to cater for rules that embed possibly nonfactual propositions. However, there does not appear to be any simple solution to providing explicit, formal semantics for such rules. As a nasty example, consider the following business rule:

It is not permitted that some department adopts a rule that says it is obligatory that each employee of that department is male.

This example includes the mention (rather than use) of an open proposition in the scope of an embedded deontic operator. One possible, though weak, solution is to rely on reserved domain predicates to carry much of the semantics implicitly. For example, the ORM schema in Figure 7 uses the special predicates "obligates the actualization of" and "is actual," as well as an object type "PossibleAllMaleState," which includes all conceivable all-male states of departments, whether actual or not. The derived fact type "PossibleAll-MaleState is actual" may be defined using the derivation rule:

PossibleAllMaleState is actual **iff**

    PossibleAllMaleState is of **a** Department **and**

    each Person **who** works for **that** Department is male, that is, $\forall x$:PossibleAllMaleState [x is actual $\equiv \exists y$:Department (x is of y & $\forall z$:Person (z works for y $\supset$ z is male))]. The deontic constraint may now be captured by the following textual constraint on the fact type "RuleAdoption is forbidden":

RuleAdoption is forbidden **if**

RuleAdoption is by **a** Department

        and is of **a** Rule

             that obligates the actualization of **a** PossibleAllMaleState

                that is of **the same** Department,

that is, $\forall x$:RuleAdoption $\forall y$:Department $\forall z$:Rule $\forall w$:PossibleAllMaleState [(x is by y & x is of z & z obligates the actualization of w & w is of y) $\supset x$ is forbidden].

The formalization of the deontic constraint works because the relevant instance of PossibleAllMaleState exists, regardless of whether or not the relevant

Figure 7.A complex case involving embedded mention of propositions

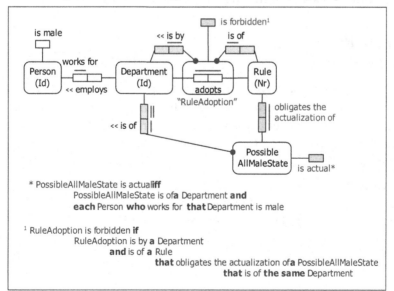

department actually is all male. The "obligates the actualization of" and "is actual" predicates embed a lot of semantics, which is left implicit. While the connection between these predicates is left informal, the derivation rule for "PossibelAllMaleState is actual" provides enough semantics to enable human readers to understand the intent.

Alternatively, we could adopt one of two extremes: (a) treat the rule overall as an uninterpreted sentence, or informal comment, for which humans are to provide the semantics, or (b) translate the semantic formulation directly into higher order logic, which permits logical formulations (which connote propositions) to be predicated over. The complexity and implementation overhead of Option 2 would seem to be very substantial.

We could try to push such cases down to first-order logic by providing the necessary semantic formulation machinery as a predefined package that may be imported into a domain model, and then identifying propositions by means of a structured logical formulation. However, that seems unclean because in order to assign formal semantics to such expressions, we must effectively adopt the higher order logic proposal mentioned in the previous paragraph.

Support for reification might be added as an extension to common logic at some future date. This support is intended to cater for objectification of propositions that are already being asserted as facts (i.e., propositions being used), as well as propositions for which no factual claim is made (i.e., propositions being mentioned), while still retaining a first-order approach. When available, this may offer a better solution for the problem under consideration.

Dynamic Rules

Dynamic constraints apply restrictions on possible transitions between business states. The constraint may simply compare one state to the next (e.g., salaries should never decrease), or the constraint may compare states separated by a given period (e.g., invoices ought to be paid within 30 days of being issued).

The invoice rule might be formally expressed in a high-level rules language, thus assuming the fact types "Invoice was issued on Date" and "Invoice is paid on Date" are included in the conceptual schema:

For each Invoice, **if that** Invoice was issued on Date₁ **then it is obligatory that that** Invoice is paid on $Date_2$ **where** $Date_2 <= Date_1 + 30$ days.

This might now be normalized to the following formulation, moving the deontic operator to the front:

It is obligatory that each Invoice **that** was issued on $Date_1$ is paid on $Date_2$ **where** $Date_2 <= Date_1 + 30$ days.

There are two issues here. First, what rules did we rely on to license the transformation of the rule? It would seem that we require an equivalence rule such as $p \supset Oq .\equiv. O(p \supset q)$. While this formula is actually illegal in some deontic logics, it does seem intuitively acceptable. At any rate, the preliminary transformation work in normalizing a business-rule formulation might involve more than just the Barcan equivalences or their deontic counterparts. In principle, this issue might be ignored for interoperability purposes so long as the business-domain expert is able to confirm that the final normalized formulation (perhaps produced manually by the business-rules modeler) agrees with their intended semantics; it is only the final, normalized formulation that is used for exchange with other software tools.

The second issue concerns the dynamic nature of the rule. While it is obvious how one may actually implement this rule in a database system, capturing the formal semantics in an appropriate logic (e.g., a temporal or dynamic logic) is a harder task. One possibility is to provide a temporal package that may be imported into a domain model in order to provide a first-order logic solution. Another possibility is to adopt a temporal modal logic (e.g., treat a possible world as a sequence of accessible states of the fact model). For a discussion of why we prefer a first-order solution where possible, see Halpin (2005a).

Conclusion

In practice, many business constraints are of a deontic rather than alethic nature. This chapter discussed an approach for adding formal support for deontic constraints within information models using ORM 2 to illustrate various examples. NORMA, an open-source ORM 2 tool, is being used as a vehicle to implement the suggested approach. Although still at the prototype stage, this tool already provides automated verbalization of alethic and deontic constraints. While the ORM 2 modeling notation was used to illustrate the ideas, the notion of adding support for deontic constraints is just as relevant for other modeling approaches such as ER and UML, and much of the formal discussion in the chapter applies equally well to these approaches.

The formalization of static constraints of both alethic and deontic modalities was discussed in some depth. NORMA's modality support is restricted to those modal formulae that include just one modal operator ("it is necessary that," "it is obligatory that"), where that operator is the main operator. Such formulae appear to offer no major implementation difficulties. However, more complex formulae involving either embedded deontic operators or embedded mention of propositions are far harder to support. While the chapter identified some possible approaches to address these complex cases, further research is needed to determine the best solution. The topic of modalities in dynamic constraints also needs further research.

Acknowledgment

Some aspects of the logical formalization presented in this chapter have benefited from discussions with Pat Hayes (IHMC, Florida).

References

Barker, R. (1990). *CASE*method: Entity relationship modelling.* Wokingham: Addison Wesley.

Chen, P. P. (1976). The entity-relationship model: Towards a unified view of data. *ACM Transactions on Database Systems, 1*(1), 9-36.

De Troyer, O., & Meersman, R. (1995). A logic framework for a semantics of object oriented data modeling. In *Proceedings of the 14th International ER Conference* (LNCS 1021, pp. 238-249). Gold Coast, Australia: Springer.

Girle, R. (2000). *Modal logics and philosophy.* McGill-Queen's University Press.

Halpin, T. (1989). *A logical analysis of information systems: Static aspects of the data-oriented perspective.* Unpublished doctoral dissertation, University of Queensland, Queensland, Australia.

Halpin, T. (2001). *Information modeling and relational databases.* San Francisco: Morgan Kaufmann.

Halpin, T. (2004a). Business rule verbalization. In A. Doroshenko, T. Halpin, S. Liddle, & H. Mayr (Eds.), *Information systems technology and its applications* (LNI P-48, pp. 39-52). Salt Lake City, UT.

Halpin, T. (2004b). Comparing metamodels for ER, ORM and UML data models. In K. Siau (Ed.), *Advanced topics in database research* (Vol. 3, pp. 23-44). Hershey, PA: Idea Group Publishing.

Halpin, T. (2005a). Higher-order types and information modeling. In K. Siau (Ed.), *Advanced topics in database research* (Vol. 4, chap. 10, pp. 218-237). Hershey, PA: Idea Group Publishing.

Halpin, T. (2005b). Objectification. In *Proceedings of CAiSE'05 Workshops* (Vol. 1, pp. 519-532).

Halpin, T. (2005c). ORM 2. In R. Meersman, Z. Tari, P. Herrero et al. (Eds.), *On the Move to Meaningful Internet Systems 2005: OTM 2005 Workshops* (LNCS 3762, pp. 676-687). Cyprus: Springer.

Halpin, T. (2006). Object-role modeling (ORM/NIAM). In P. Bernus, K. Mertins, & G. Schmid (Eds.), *Handbook on architectures of information systems* (2nd ed., pp. 81-103). Berlin, Germany: Springer-Verlag.

ISO. (2005). *ISO common logic standard* (Draft). Retrieved from http://cl.tamu.edu/docs/cl/32N1377T-FCD24707.pdf.

Krogstie, J., & Sindre, G. (1996). Utilizing deontic operators in information system specification. *Requirements Engineering Journal, 1*, 210-237.

Object Management Group (OMG). (2003a). *UML 2.0 infrastructure specification.* Retrieved from http://www.omg.org/uml

Object Management Group (OMG). (2003b). *UML 2.0 superstructure specification.* Retrieved from http://www.omg.org/uml

Object Management Group (OMG). (2006). *Semantics of business vocabulary and rules interim specification.* Retrieved from http://www.omg.org/cgi-bin/doc?dtc/06-03-02

Rumbaugh, J., Jacobson, I., & Booch, G. (1999). *The unified language reference manual.* Reading, MA: Addison-Wesley.

ter Hofstede, A. H. M., Proper, H. A., & Weide, th. P. van der. (1993). Formal definition of a conceptual language for the description and manipulation of information models. *Information Systems, 18*(7), 489-523.

Chapter IX

Lost in Business Process Model Translations:
How a Structured Approach Helps to Identify Conceptual Mismatch

Jan Recker, Queensland University of Technology, Australia

Jan Mendling,
Vienna University of Economics and Business Administration,
Austria

Abstract

Often, different process models are employed in different phases of the BPM life cycle, each providing a different approach for capturing business processes. Efforts have been undertaken to overcome the disintegration of process models by providing complementary standards for design and execution. However, this claim has not yet been fulfilled. A prominent example is the seemingly complementary nature of BPMN and BPEL. The mapping between these process modeling languages is still unsolved and poses challenges to practitioners and academics. This chapter discusses the problem of translating between process modeling languages. We argue that there is conceptual mismatch between modeling languages stemming from various perspectives of the business-process management life cycle that must be identified for seamless integration. While we focus on the popular case of BPMN vs. BPEL, our approach is generic and can be utilized as a guiding framework for identifying conceptual mismatch between other process modeling languages.

Introduction

Business process models play a key role in both organizational management (Davenport & Short, 1990; Hammer & Champy, 1993; Smith & Fingar, 2003) and information systems development (Curtis, Kellner, & Over, 1992; Dumas, van der Aalst, & ter Hofstede, 2005; Ellison & McGrath, 1998). In theory, business-process modeling (BPM) efforts follow a certain life cycle (Smith & Fingar; Weske, van der Aalst, & Verbeek, 2004; zur Muehlen, 2004) that idealizes the phases of development and deployment of business processes into the stages of design, implementation, enactment, and evaluation.

In principle, the design phase involves the development of conceptual process models from a business analyst perspective. During this phase, business processes are documented in an intuitive form to communicate the business requirements to relevant stakeholders. In a second step, these models serve as input to technical analysts concerned with the development of technical process models, that is, implementation models in the form of executable work-flow specifications. These specifications then serve as templates for the enactment of process instances deployed on work-flow engines. Lastly, the execution of a process is monitored and evaluated by process controlling and analysis tools to guide the revision and improvement of the process models as part of another iteration of the life cycle.

While in theory the business-process life cycle proposes a seamless interplay between the various phases, in business practice the transition between the phases is often broken. For instance, a wide range of different process modeling languages can be employed in the various stages of the life cycle, each with a different focus on audience and modeling purpose (Bider & Johannesson, 2002; Katzenstein & Lerch, 2000). Some of the languages provide mechanisms to develop high-level conceptual models that provide an understanding of an organization from an intentional and social perspective or for reasoning support during redesign (Yu, Mylopoulos, & Lespérance, 1996). Other languages provide capacities to develop lower level technical models that are especially suited for the description, execution, and simulation of business processes. Not surprisingly, the process design and execution stages indeed usually employ different process modeling languages, and in effect the translation between the languages is prone to semantic ambiguities (zur Muehlen & Rosemann, 2004). This may in turn cause the loss of design considerations within the execution models.

We refer to such undesirable cases as conceptual mismatch between process modeling languages deployed in different phases of the BPM life cycle. Accordingly, the transition between the phases seems to be an important prerequisite to make the BPM life cycle work, in particular, between business analyst and technical analyst models (Dreiling, Rosemann, & van der Aalst, 2005; zur Muehlen & Rosemann, 2004). Related efforts have repeatedly tried to overcome the gap between the process model life cycle phases and to bridge business models with technical process specifications, for example, Dehnert and van der Aalst (2004).

The most recent and popular example for work that addresses this transition is the case of the recently proposed Business Process Modeling Notation (BPMN; BPMI.org & Object Management Group [OMG], 2006). BPMN has been developed to enable business analysts to develop readily understandable graphical representations of business processes and to enable technical analysts to represent complex process semantics. Its developers specifically claim that BPMN is supported with appropriate graphical object properties that will enable the generation of executable work-flow models that comply with the BPEL (business process execution language) specification (Andrews et al., 2003). This would indeed bridge the gap between business analyst and technical analyst perspectives by providing a standard visual notation for executable processes. In fact, the specification document states that "BPMN creates a standardized bridge for the gap between the business process design and process implementation" (BPMI.org & OMG, p. 1). However, as we will discuss during the course of this chapter, the translation of BPMN to BPEL is far from trivial.

In this chapter we show that mapping issues arise foremost from conceptual mismatch that exists between process modeling languages. This argument is based on the observation that languages, in their essence, differ in expressive power, which in turn hinders the translation of models between the languages.

Accordingly, the first and foremost objective of this chapter is to discuss how conceptual mismatch between business analyst and technical analyst process models can be identified. Despite the focus on BPMN and BPEL, we seek to deliver a generic solution that builds on established evaluation theories in the field of process modeling. Forthcoming from this discussion, as a second contribution of this chapter we provide guidance for the translation of process models in the form of abstract transformation strategies that we deem promising for overcoming the identified mismatch.

We proceed as follows. In the next section we briefly introduce our selected example languages, BPEL and BPMN, to give the reader sufficient background for understanding our subsequent elaborations. Also, we discuss existing related studies on the correspondence between process modeling languages, which will show that there indeed is significant mismatch between process modeling languages that hinders if not counteracts translation specifications. Following the background section we then derive a multiperspective approach for identifying conceptual mismatch between business process modeling languages and apply it to BPMN and BPEL. We then close in the last section by drawing some conclusions from our work and by discussing some future trends.

Background

In the remainder of this chapter we will repeatedly refer to the languages of BPMN and BPEL, and also recapitulate previous analyses of these languages that were conducted in preparation of this study. In this section we thus briefly introduce BPMN and BPEL in order to enable the reader to comprehensively follow our elaborations later on.

BPMN

Across their life cycle, process models in general serve two main purposes. During the initial stages, intuitive business-process models are used for scoping the project, and capturing and discussing business requirements and process-improvement initiatives with subject-matter experts. A prominent example of a business modeling technique used for such purposes is the event-driven process chain (Keller, Nüttgens, & Scheer, 1992). At later stages of the life cycle, business-process models are used for process automation, which requires their conversion into executable specifications. Techniques used for depicting process models for this purpose have higher requirements in terms of expressive power. Examples include Petri nets (Petri, 1962) or YAWL (yet another workflow language; van der Aalst & ter Hofstede, 2005).

However, the nature of these technical and/or executable process description languages renders them less suited for direct use by nonexperts in order to design, manage, and monitor the business processes that are enacted by process-aware information systems. On the other hand, many of the intuitive process modeling languages do not provide sufficient support for more technical-oriented purposes such as simulation or execution.

Clearly, what is needed is a standard visual notation for business processes that is both intuitive and also supportive of process execution. The Business Process Management Initiative (BPMI; www.bpmi.org) recognized this need and started work on the Business Process Modeling Notation in early 2003. Version 1.0 of BPMN was first released in May 2004 and in February 2006 was approved by the OMG as a final adopted specification (BPMI.org & OMG, 2006) for standardization purposes.

The development of BPMN was driven by two objectives: on the one hand to develop a modeling language that supports typical process modeling activities both for business and technical users, and on the other hand to provide a standard visualization mechanism for executable process specifications (essentially, for BPEL processes) that also supports the automatic mapping from BPMN models to BPEL specifications.

The complete BPMN specification defines 38 distinct language constructs plus attributes, grouped into four basic categories of elements, each of which will briefly be introduced in the following:

- **Flow objects:** Flow objects are the main graphical elements used to create business-process diagrams (BPDs). They define the behavior of a business process by means of objects such as events, activities, and gateways.
- **Connecting objects:** Connecting objects are used to connect flow objects through different types of arcs to each other or to other information. They can be sequence flows, message flows, or association flows.
- **Swimlanes:** Swimlanes are used to group activities into separate categories for different functional capabilities or responsibilities (e.g., different

roles or organizational departments). There are two ways of grouping the primary modeling elements through swimlanes: either via pools or via lanes.

*   **Artifacts:** Artifacts are used to provide additional information about the process, such as processed data or other comments. Currently, BPMN supports the artifacts data object, group, and annotation.

Figure 1 gives an example of a simple business-process diagram depicted in BPMN. A business process of a retailer is executed by the sales and the distribution departments, each of which is represented as a separate lane in the Retailer pool. The credit card authentication activity involves an interaction with a financial institution, depicted as a separate pool. The different activities are depicted as rounded boxes connected with control-flow arcs. Gateways are, for instance, used to define decision points. Moreover, the process of each process participant starts with a start event and terminates with an end event.

For further details on BPMN, refer to the specification (BPMI.org & OMG, 2006).

Since its initial publication (BPMI.org, 2004), BPMN has been accepted by a large part of the BPM community, predominantly due to the claim of mapping directly to executable process languages including XPDL (Fischer, 2005) and

Figure 1. A simple business-process diagram

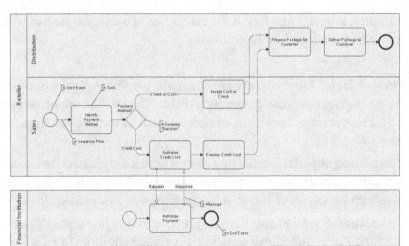

BPEL (Andrews et al., 2003). The wide uptake of the notation by most BPM tool vendors (BPMI.org, 2005) further indicates a high potential for longevity. Some practitioners have hailed BPMN as supplying a rich representation that allows business-process management systems the ability to control the required interactions with humans and third-party applications in the design phase (Miers, 2003). Furthermore, analyses of BPMN from analytical (e.g., Wohed, van der Aalst, Dumas, ter Hofstede, & Russell, 2006) and empirical perspectives (e.g., Nysetvold & Krogstie, 2005; Recker, Indulska, Rosemann, & Green, 2006) confirm its considerable level of sophistication in representing concepts required for modeling business processes.

BPEL

The business process execution language for Web services (Andrews et al., 2003) is, in its essence, an extension of imperative programming languages with constructs specific to the BPM domain, in particular Web service implementations. Version 1.1 of BPEL was released in 2003 and its Version 2.0 is currently in the process of standardization with OASIS. A BPEL process definition specifies the technical details of a work flow that offers a complex Web service built from a set of elementary Web services.

Six of BPEL's most important concepts are briefly presented in the following, that is, partner links, variables, correlation, basic activities, structured activities, and handlers.

- **Partner links:** A partner link provides a communication channel to a remote Web service to be used in the BPEL process. A respective partner link type must be defined first to specify the required and provided WSDL port types.

- **Variables:** Variables are used to store both message data of Web service interactions and control data of the process. A variable must be declared in the header of a BPEL process by referencing a WSDL or an XML (extensible markup language) schema data type.

- **Correlation:** As BPEL supports long-running business processes, there may be several process instances waiting for Web service messages at a certain point of time. A correlation set specifies so-called properties, that is, queries to retrieve message parts that are unique for a specific process instance. According to a certain property value, like, for instance, ordernumber = 1002007, a message is handed to the matching process instance.

- **Basic activities:** The elementary steps of a BPEL process are performed by basic activities. There are activities to send and receive messages from Web services (receive, invoke, reply), to change the content of variables (assign), to wait for a certain period or up to a certain point in time (wait), and to terminate the process (terminate). The upcoming second, revised version of BPEL will introduce an activity to check conformance to a schema (validate) and the possibility to add proprietary activities (extensionActivity).

- **Structured activities:** The control flow of basic activities can be defined in two different styles: block oriented or graph based. Both styles can be mixed. Block-oriented control flow can be defined with structured activities. BPEL offers activities to specify parallel execution (flow), conditional branching based on data (switch) or on receipt of a message (pick), and sequential execution (sequence). Structured activities can be nested. Scopes are special structured activities. They demarcate the scope of local variables and handlers. Control flow can also be defined as graph based, but without introducing cycles, using so-called links. A link represents a synchronization between two activities.

- **Handlers:** BPEL provides handlers to deal with unexpected or exceptional situations. Event handlers wait for messages or time events. They can be used to specify deadlines on the process level. Fault handlers catch internal faults of the BPEL process. If the fault cannot be resolved, the compensation handler can be triggered to undo the effects of already completed activities. Finally, the termination handler to be introduced in BPEL 2 will offer a mechanism to force a process to terminate, for example, due to external faults.

Even though BPEL supports a rich set of primitives to specify executable processes, there are still some features missing toward a full-fledged business-process specification. The extension activity of BPEL 2 is a useful anchor point to fill these gaps. Currently, there are several BPEL extensions in progress of development, in particular, BPELJ for Java in-line code, BPEL4People for human work lists (both available from ftp://www6.software.ibm.com/software/developer/library/), and BPEL-SPE for subprocesses (Kloppmann et al., 2005). For further details on BPEL, refer to the specification (Andrews et al., 2003).

On the Correspondence Between BPMN and BPEL

The transition of process models between the various stages of the BPM life cycle has been posing research questions for quite some time. However, most of the previous failed to achieve satisfactory solutions that were able to gain widespread acceptance in process modeling and management practice. The recent momentum of BPMN and BPEL in the industry has further triggered related research to study the correspondence between process modeling languages. In this section we briefly recapitulate work on the correspondence between process modeling languages, again using the example of the BPMN-BPEL case.

Trying to support the claim that BPMN provides a visualization notation for BPEL, subsection 11 of the BPMN specification (BPMI.org & OMG, 2006, pp. 137-204) presents a mapping from BPMN to BPEL. However, it is rather informally given in prose; a precise algorithm and a definition of required structural properties are missing. An example of how a mapping could work is given in White (2005), but it is rather simple and the feasibility of such a mapping in the general case has not been demonstrated yet. Other examples of how to use BPMN to model BPEL processes are also given in White. Again, however, they do not reach levels of process complexity that can be considered realistic. The same unfortunately holds for the proposed mapping from UML (unified modeling language) activity diagrams to BPEL (Mantell, 2005) that fails to address some more difficult process modeling scenarios. It is further worthwhile noting that some available software such as Telelogic's System Architect (http://www.telelogic.com/popkin/) support the generation of BPEL code from BPMN diagrams, but only for a limited subset of BPMN.

From an academic perspective, recent work has led to the proposal of transformation strategies for process models, with focus often given to the case of BPMN and BPEL. Ouyang, Dumas, Breutel, and ter Hofstede (2006) present a general approach to translate standard work-flow models—an abstraction of a set of process modeling languages, such as, for instance, BPMN and UML activity diagrams, to an arbitrary topology of elementary work-flow constructs (Kiepuszewski, ter Hofstede, & van der Aalst, 2003)—to BPEL by exploiting the BPEL construct of event handler. However, as the authors admit, this approach only holds for a core subset of BPMN and UML activity diagrams. Later, this approach was adopted for the specific context of BPMN and BPEL (Ouyang, van der Aalst, Dumas, & ter Hofstede, 2006). Again, the approach exploits BPEL event handlers for unstructured subsets of the BPMN models whilst also defining a translation algorithm that is capable of generating readable BPEL code by discovering certain patterns in BPMN models that can be mapped onto BPEL structured constructs. While this approach, too, is not yet at a stage where it holds for more advanced BPMN models, it is closely related to our forthcoming discussion as we also take into account the mismatch between BPMN and BPEL with respect to the support for patterns in the control-flow representation of process modeling languages. Gao (2006) presents an approach using two-phase transformations of BPMN diagrams to BPEL specifications. Again we observe a missing proof of general feasibility and applicability. Another interesting approach is discussed in Mendling, Lassen, and Zdun (2006), where the authors discuss different strategies for translating graph-oriented models (like BPMN) to block-oriented specifications (like BPEL). These strategies have different perks and perils; nevertheless, we deem them a suitable starting point for devising concrete mappings based on the identification and understanding of the mismatch between the languages. Hence, we will refer back to them later in this chapter.

Conceptual Mismatch Between Process Modeling Languages

As the discussion of related work reveals, existing transformation strategies between process modeling languages regularly falter when it comes to defining general mappings. We argue that the root cause for such translation problems resides in the conceptual mismatch that exists between any two process modeling languages. We make two observations in the context of BPMN and BPEL to exemplify this argument.

First, BPEL and BPMN come from different backgrounds (technical analyst vs. business analyst). Thus, they employ different paradigms for capturing relevant aspects of business processes, which in turn leads to the manifestation of conceptual mismatch with respect to the expressive power of these languages. Second, BPEL and BPMN are usually employed in different stages of the BPM life cycle. Hence, the requirements of both stages need to be taken into consideration when identifying potential conceptual mismatch.

Based on these observations, we argue that the different BPM life-cycle perspectives need to be taken into consideration when devising a transformation between process models. Specifically, we argue that there are three perspectives to consider.

From a business analyst perspective, the transition between process modeling languages such as BPMN and BPEL must preserve the semantic information about the represented domain; that is, it should minimize if not avoid loss of semantic representation information. In this regard, Wand and Weber's (1990, 1993, 1995) work is widely acknowledged as a framework of real-world domain concepts that modeling languages should be able to represent. In other words, a transition between languages should establish a high extent of matching domain representation capabilities between the two languages.

From a technical analyst perspective, the underlying work-flow execution engine determines the specification of processes. In this regard, Kiepuszewski et al. (2003) state that control flow is a central aspect of a business process that needs to be sufficiently supported by any given language or work-flow execution engine. Therefore, a transition between languages should establish a high extent of matching support for various aspects of control flow.

Beyond these life-cycle-specific perspectives, a more general observation must be made with respect to the process representation paradigm that underlies any process modeling language. Different paradigms provide different lenses through which a process is conceptualized and ultimately depicted in a process modeling language. The most common process representation paradigms are block-oriented vs. graph-oriented process representations (Mendling et al., 2006; Ouyang, Dumas, et al., 2006). We argue that different process representation paradigms potentially denote another source of conceptual mismatch between process modeling languages. While both domain representation capabilities and control-flow support permit statements about what types of relevant aspects of a process can be expressed, the process representation paradigm influences how such aspects can be expressed.

Forthcoming from these argumentations, an approach for identifying the conceptual mismatch between business and technical analyst process models must be able to identify all three types of conceptual mismatch. We will employ two established evaluation frameworks, namely, representation theory (Wand & Weber, 1990, 1993, 1995) for the specification of domain representation capability mismatch, and the workflow patterns framework (van der Aalst, ter Hofstede, Kiepuszewski, & Barros, 2003) for the specification of control-flow support mismatch. In addition to these established theories, we also introduce a mismatch identification method with respect to the process representation paradigms employed, based on a set of transformation strategies (Mendling et al., 2006) that can potentially be used to translate process models into another.

The selection of the mentioned evaluation frameworks can be reasoned by their levels of maturity, rigorous development, and structured evaluation approach as well as by their established track record in the field of process modeling. For overviews refer, for instance, to Rosemann, Recker, Indulska, and Green (2006) and Wohed et al. (2006), respectively. In particular, as we seek to deliver a general contribution beyond the case of BPMN and BPEL, the high level of dissemination of these theories in the field of process modeling reasons our selection as it allows for a wider uptake of our approach to cases of other process modeling languages that have previously been evaluated, such as, for instance, BPML and WSCI (Green, Rosemann, Indulska, & Manning, in press; van der Aalst, Dumas, ter Hofstede, & Wohed, 2002).

Identifying Domain Representation Capability Mismatch

Conceptual modeling languages, such as process modeling languages, in their essence are used to build a representation of selected phenomena in the problem domain for the purpose of understanding and communication among stakeholders (Kung & Sølvberg, 1986; Mylopoulos, 1992; Siau, 2004). As such, an important criterion for process modeling languages is their capability to develop good descriptions of the real-world domains that the modeler seeks to capture in the process model. A good description embraces the notion of completeness as an indication of what types of real-world phenomena a process modeling language is able to represent.

Over the last decades, models of representation such as representation theory proposed by Wand and Weber (1990, 1993, 1995) have increasingly been used as a theoretical reference benchmark to assess the completeness of process modeling languages, that is, their capabilities to depict all relevant real-world phenomena in a model (Rosemann et al., 2006).

These models of representation, for instance, the well-known Bunge-Wand-Weber (BWW) representation model (Wand & Weber, 1990, 1993, 1995; Weber, 1997), are based on theories of ontology. Ontology is a well-established theoretical domain within philosophy dealing with identifying and understanding elements of the real world (Bunge, 2003). Today, however, interest in, and the applicability of, ontologies extends to areas well outside of philosophy (see, e.g., Gruber, 1993; Uschold & Grüninger, 1996). Especially in the area of conceptual modeling for information systems analysis and design have ontologies emerged as fruitful theoretical bases on which to establish concepts and phenomena associated with modeling real-world domains (Green & Rosemann, 2004; Guizzardi, 2005; Milton & Kazmierczak, 2004).

The BWW representation model as the most popular and widely used instance of representation theories has over recent years achieved significant levels of scholarly attention and dissemination. It is documented by well over 100 publications drawing on this model in contexts such as the comparison of modeling languages (Rosemann et al., 2006), modeling language foundations (Wand, Monarchi, Parsons, & Woo, 1995), model quality measurement (Gemino & Wand, 2005), and modeling method engineering (Wand, 1996). It specifies a number of representation constructs that are deemed necessary to faithfully provide complete and clear representations of information systems domains. We omit a more in-depth discussion of the model and its previous applications to the area of process modeling in this chapter and instead refer the reader to the overview given, for instance, in Rosemann et al. (2006).

Representation theory prescribes a procedure for evaluating modeling languages as to their capability to express various aspects of real-world domains, known as representational analysis (Recker et al., 2006). During this process, the constructs of the BWW representation model (e.g., thing, state, transformation) are compared with the language constructs of the modeling language (e.g., event, activity, actor). Amongst other aspects, this comparison reveals construct deficit within a modeling language, that is, the extent to which a modeling language has a deficit of constructs mapping to the set of constructs proposed in the BWW representation model. This in turn serves as an indication that the modeling language under observation is limited in its capacity to make statements about all relevant phenomena of real-world domains (Weber, 1997).

Whilst the investigation of construct deficit within a given language allows for conclusions on the scope of coverage of the respective language, we are here interested in finding out whether any two languages share the same extent of deficit, that is, only those types of construct deficit in a particular language (e.g., BPMN) that another language (e.g., BPEL) is able to express. We argue that this particular form of deficit constitutes a form of mismatch that in turn potentially impacts the translation of models between these languages. This means that if a more expressive process modeling language features a representation construct that is not supported in a less expressive process modeling language, then the translation of the modeled process to the less detailed language will be at the cost of losing expressive power and thus semantic information about the represented domain.

For the purpose of this chapter, we draw on the individual analyses of BPMN (Recker et al., 2005, 2006) and BPEL (Green et al., in press) that were conducted in preparation for this study. Rosemann and Green (2002) showed that the BWW model can be represented in a metamodel that shows several clusters of BWW constructs: things including properties and types of things, states assumed by things, events and transformations occurring on things, and systems structured around things. We use this proposed clustering to structure our line of investigation of the differences between BPMN and BPEL in terms of their construct deficit (see Table 1). It must be noted that representation theory offers a systematic analytical method called overlap analysis (Green et al., in press; Weber, 1997) for a thorough and more detailed evaluation of the completeness and overlap of domain representations in any combination of languages. We must consider such an evaluation out of the scope of this chapter. Yet, we see an interesting and important research challenge in such an overlap analysis in order to more comprehensively and

Table 1. Support for the BWW model constructs in BPMN and BPEL (Adapted from Green et al., in press; Recker et al., 2005, 2006)

BWW Construct	Cluster	BPMN	BPEL
THING		++	-
PROPERTY		*N/A*	*N/A*
In General		++	+
In Particular		-	-
Hereditary		-	-
Emergent	Things including properties and types of things	-	+
Intrinsic		-	-
Nonbinding Mutual		-	-
Binding Mutual		-	-
Attributes		-	+
CLASS		++	+
KIND		+	-

continued on following page

Table 1. continued

STATE	States assumed by things	-	+
CONCEIVABLE STATE SPACE		-	-
LAWFUL STATE SPACE		-	-
STATE LAW		-	-
STABLE STATE		-	-
UNSTABLE STATE		-	-
HISTORY		-	-
EVENT	Events and transformations occurring on things	++	++
CONCEIVABLE EVENT SPACE		-	-
LAWFUL EVENT SPACE		-	-
EXTERNAL EVENT		++	+
INTERNAL EVENT		++	++
WELL-DEFINED EVENT		++	+
POORLY DEFINED EVENT		++	++
TRANSFORMATION		++	++
LAWFUL TRANSFORMATION		++	++
Stability Condition		++	+
Corrective Action		++	-
ACTS ON		+	+
COUPLING		+	+
SYSTEM	Systems structured around things	++	+
SYSTEM COMPOSITION		++	+
SYSTEM ENVIRONMENT		++	-
SYSTEM STRUCTURE		-	+
SUBSYSTEM		++	-
SYSTEM DECOMPOSITION		++	-
LEVEL STRUCTURE		++	-

rigorously clarify the type of mismatch between the combination of any two process modeling languages.

Table 1 summarizes the findings from the analyses in Green et al. (in press) and Recker et al. (2005, 2006). In this table, a "+" indicates that the respective language provides one construct supporting the representation of the respective BWW model construct, a "++" indicates a support for the BWW model construct by more than one language construct, and a "–" indicates a lack of support for the respective BWW model construct.

As can be seen from Table 1, there are a number of potential domain representation capability mismatches between BPMN and BPEL, indicated by varying support for the BWW model constructs. The following paragraphs discuss some of these discrepancies with respect to a potential translation of process models from BPMN to BPEL.

Translation of Things Including Types and Properties of Things

A thing denotes the elementary notion in representation theory (Weber, 1997). The perceived world is constituted of things, either imaginary or real, that can be grouped into sets and species of things (class and kind, respectively). Table 1 reveals that BPMN is capable of representing things, classes, and kinds of things. However, BPEL only supports the representation of classes of things; that is, BPEL can only make semantic statements about groups of things but not specific instances. This means that object instances in a BPD, for example, a specific organizational entity, a specific business partner, or a specific application system, possibly need to be generalized to classes of instances, that is, to a more aggregate level. In this regard, Bodart, Patel, Sim, and Weber (2001) point out that the use of optionality that stems from a focus on classes, as opposed to instances or subtypes, may result in a superficial understanding of the specification.

On the other hand, the rather limited and general representation of properties of things in BPMN can be broken down into more specialized subtypes of properties in BPEL.

Translation of States Assumed by Things

A state of a thing is a vector of all the property values of a thing at a given point of time (Weber, 1997). Table 1 reveals that both BPMN and BPEL lack expressive power for modeling states assumed by things. While this finding may be problematic in general—see also the discussions in Rosemann et al. (2006) and the related findings in Green and Rosemann (2000) and Recker et al. (2006)—it does not denote an area of concern with respect to translating BPMN diagrams to BPEL as both languages basically share the same incapability for explicit state representation.

Translation of Events and Transformations Occurring on Things

The occurrence of an event changes the state of a thing. A transformation is the mapping between two states of a thing (Weber, 1997). Table 1 reveals that BPMN has more expressive power than BPEL for the representation of events and transformations occurring on things. As an example, BPMN offers constructs to specify corrective actions and stability conditions to determine transformations that are lawful. Corresponding concepts in BPEL seem to be specified in an implicit manner in exception handling and compensation activities rather than explicitly in dedicated representation constructs.

Generally, there seems to be a high extent of redundancy of BPMN in terms of transformation and event modeling (Recker et al., 2006); that is, BPMN offers many overlapping constructs and thus lacks orthogonality. A translation to BPEL potentially needs to map certain dedicated event subtypes within BPMN to a single event type of BPEL (for instance, an external event). Transformations, on the other hand, are more differentiated in BPEL. This implies that representations of transformations in BPMN potentially need to be annotated with further information or attributes to sufficiently specify a mapping to an appropriate BPEL construct.

Translation of Systems Structured Around Things

Things can be composed to a system, which may have subsystems and inter-faces to the environment of the system (Weber, 1997). Table 1 reveals that BPMN's support for the modeling of systems structured around things excels the support provided by BPEL. Thus, a BPMN specification of the system to be developed, especially the demarcation from its environment (system environment) and its disaggregation into subsystems (system decomposition), may not be unambiguously translatable into executable BPEL specifications and may thus require extra modeling and specification effort to avoid misin-terpretations of the resulting BPEL models. In particular, the mapping of the BPMN pool and lane constructs to the BPEL partner construct will require attention as the semantics of pool and lane seem to be more extensive than their BPEL counterpart.

Identifying Control-Flow Support Mismatch

Given the objectives of the later, more technical stages of the process model life cycle, that is, specifying processes for execution, there are requirements for process modeling languages to support various aspects of work flow that are being enacted in any given work-flow execution engine. In order to be able to identify which aspects of control flow are supported by leading work-flow management systems and to evaluate which of the given process modeling languages are able to match the requirements of these systems, the workflow patterns framework was developed. More precisely, the development of the workflow patterns framework (http://www.workflowpatterns.com) was triggered by a bottom-up analysis and comparison of different leading work-flow management software. The goal was to bring insights into the expressive power and capabilities of the underlying work-flow and business process modeling languages. The framework consists of a number of patterns and provides a taxonomy of generic, recurring concepts and constructs relevant in the context of process automation, simulation, and execution. In accordance with Jablonski and Bussler's (1996) original classification, the framework was gradually developed to cover the control flow, the data, and the resource perspectives, and it incorporates 20 control-flow patterns (van der Aalst et al., 2003), 43 resource patterns (Russell, van der Aalst, ter Hofstede, & Edmond, 2005), and 40 data patterns (Russell, ter Hofstede, Edmond, & van der Aalst, 2005). We will here focus on the control flow perspective in which six clusters can be identified that specify atomic chunks of behavior capturing some specific process control requirements. Basic control-flow patterns define the basic aspects of process control. Advanced synchronization patterns define evolved but still generic control-flow scenarios that are relatively common in business-process scenarios but only scarcely supported in the earlier generations of process modeling languages. Structural patterns identify constructs that have impact on the structure of processes. Multiple-instances patterns capture behavior chunks where multiple instances of a task or activity can be created and executed simultaneously within the context of one and the same case. State-based patterns depict situations that utilize the notion of the state. Finally, cancellation patterns capture cancellation notions relevant in business scenarios.

The provided taxonomy has widely been used as a benchmark for analysis and comparison of process specification and execution languages. A comprehensive overview is, for instance, given in Wohed et al. (2006).

As with the BWW representation model, we use the workflow patterns framework and related analyses to identify the mismatch in the support for various aspects of control flow. We draw on the individual analyses of BPMN (Wohed et al., 2006) and BPEL (Wohed, van der Aalst, Dumas, & ter Hofstede, 2003) that, similar to the evaluations of BPMN and BPEL by means of representation theory, were conducted in preparation for this study. Table 2 summarizes the findings from both analyses. In this table, a "+" indicates a direct support for a pattern, a "+/–" indicates a partial support, and a "–" indicates a lack of support.

Table 2 reveals a number of mismatches between BPMN and BPEL with regard to the support for various aspects of control flow. The following paragraphs discuss some of these discrepancies, again in a cluster-oriented manner, with respect to a potential translation of process models from BPMN to BPEL.

Table 2. Support for the control-flow patterns in BPMN and BPEL (Adapted from Wohed et al., 2003; Wohed et al., 2006)

Work-Flow Patterns	Cluster	BPMN	BPEL
1. Sequence	Basic control flow	+	+
2. Parallel Split		+	+
3. Synchronization		+	+
4. Exclusive Choice		+	+
5. Simple Merge		+	+
6. Multiple Choice	Advanced syn-chro-nization	+	+
7. Synchronizing Merge		+/-	+
8. Multiple Merge		+	-
9. Discriminator		+	-
10. Arbitrary Cycles	Structural patterns	+	-
11. Implicit Termination		+	+
12. MI without Synchronization	Multiple-instances patterns	+	+
13. MI with a Priori Design-Time Knowledge		+	+
14. MI with a Priori Run-Time Knowledge		+	-
15. MI without a Priori Run-Time Knowledge		-	-
16. Deferred Choice	State-based patterns	+	+
17. Interleaved Parallel Routing		+/-	+/-
18. Milestone		-	-
19. Cancel Activity	Cancellation patterns	+	+
20. Cancel Case		+	+

Translation of Basic Control-Flow, State-Based, and Cancellation Patterns

Table 2 reveals that BPMN and BPEL both support Patterns 1 to 5 and 16 to 20 in the same manner. This means that the representations of these control-flow patterns in BPMN should be unambiguously translatable to BPEL. This finding supports the approach taken in Ouyang, Dumas, et al. (2006) and Ouyang, van der Aalst, et al. (2006), in which a mapping between BPMN and BPEL is defined based on the support for various control-flow patterns.

Translation of Advanced Synchronization Patterns

Table 2 reveals that BPMN provides almost full support for Patterns 6 to 9. BPEL, however, lacks support for multiple merges and discriminators. In particular, BPEL does not support the invocation of subprocesses (Wohed et al., 2003), which can be supported by BPMN. A specific problem is BPEL's missing support for the discriminator pattern, that is, points in the work-flow process that wait for one of the incoming branches to complete before activating the subsequent activity. Hence, discriminators used in BPMN require considerable effort in translating them to statements that are expressible in BPEL and bear the same semantics with respect to the handling of control flow.

Translation of Structural Patterns

Table 2 reveals that BPEL does not support arbitrary cycles. The While activity can only capture structured cycles, that is, loops with one entry point and one exit point. Again, this is a potential area of concern when translating arbitrary cycles from BPMN to BPEL code with equivalent control-flow semantics.

Translation of Multiple-Instances Patterns

Table 2 reveals that BPMN and BPEL both support Patterns 12, 13, and 15 in the same manner but not Pattern 14. This means that the BPMN representation of a work flow with multiple instances (where a number of instances of a given activity are initiated, and these instances are later synchronized before proceeding with the rest of the process) needs to be translated into a less expressive form in BPEL and hence, some desired control-flow support and design considerations for the modeled process are prone to getting lost.

Identifying Process Representation Paradigm Mismatch

We argue that a transformation of models must consider not only representational capabilities and control-flow pattern support, but also the underlying process representation paradigm. In this context, there are essentially two paradigms to depict processes in a process modeling language: graph-oriented and block-oriented representation (Mendling et al., 2006; Ouyang, Dumas, et al., 2006).

BPMN follows a graph-oriented paradigm using arcs to define a partial order of activities and gateways to express split and join behavior. BPEL utilizes a block-oriented paradigm to express control flow via nested structured activities enhanced with some restricted graph concepts: In a BPEL process, arbitrary synchronization can be expressed with links as long as these links are acyclic. Cycles are only allowed if they are modeled as structured loops using the While activity.

Table 3. Transformation strategies and applicable models

Transformation Strategies from BPMN to BPEL			
	Structured BPMN	*Acyclic BPMN*	*All BPMN*
Element-Preservation	-	+	-
Element-Minimization	-	+	-
Structure-Identification	+	-	-
Structure-Minimization	+	+	-
Transformation Strategies from BPEL to BPMN			
	Structured BPEL	*All BPEL*	
Flattening	+	+	
Hierarchy-Preservation	+	-	
Hierarchy-Maximization	+	+	

In Mendling et al. (2006), graph-based languages like BPMN and languages similar to BPEL are abstracted to so-called process graphs and BPEL control flow, respectively, in order to identify transformation strategies and constraints for the application of these strategies.

In this context, a process graph is called structured if split gateways match a join of the same type, and if loops are entered at one XOR join and exited at one XOR split. Furthermore, a process graph is acyclic if no node can be reached from itself. A BPEL process is structured if it does not include any links. Some transformation strategies are only applicable for process models that fulfill certain properties (see Table 3). For a formal definition of structured and cyclic process graphs as well as structured BPEL control flow, refer to Mendling et al. This reference also defines algorithms for each of the four transformation strategies that will be sketched in the following.

Transformation Strategies from BPMN to BPEL

All four transformation strategies (see Table 3) require all cycles of the BPMN process model to be structured loops with an entering XOR join and an exiting XOR split.

The idea of the element-preservation strategy is to map all BPMN elements to suitable BPEL elements nested in a BPEL flow and to define control flow with links. Gateways are mapped to BPEL empty activities that serve as targets and sources for multiple input (join) or output links (split). Since such links in a BPEL flow have to be acyclic, the BPMN model has to be without cycles, too. The advantage of this strategy is that it can be implemented quite easily and that the resulting BPEL is very similar to the graph structure of the BPMN model. Yet, as a drawback, the BPEL model has more elements (empty activities) than necessary.

The element-minimization strategy takes the result of the element-preservation strategy and replaces the empty activities with links containing transition conditions. The BPMN model also has to be acyclic. This results in the benefit of less elements, but on the other hand, it becomes more difficult to identify the correspondences between the BPMN and BPEL model.

The structure-identification strategy works similar to the transformation proposed in the BPMN specification (BPMI.org & OMG, 2006). Structured blocks can be identified via graph reduction rules defined in Mendling et al. (2006). It is an advantage that the resulting BPEL code is easy to read. Still, this strategy is only applicable if all control flow can be mapped to BPEL structured activities.

If not, the structure-maximization strategy can be applied to derive a BPEL process with as many structured activities as possible nested in a flow for additional synchronization constraints. This strategy can be applied as long as all loops can be mapped to a BPEL While. Yet, it is a drawback that the implementation of this strategy requires the most effort.

As Table 3 shows, there is no strategy to generate BPEL from an arbitrary BPMN graph because BPEL does not permit the modeling of arbitrary cycles. If the BPMN graph is structured, the structure-identification strategy can

Figure 2. Transformation strategies from BPMN to BPEL

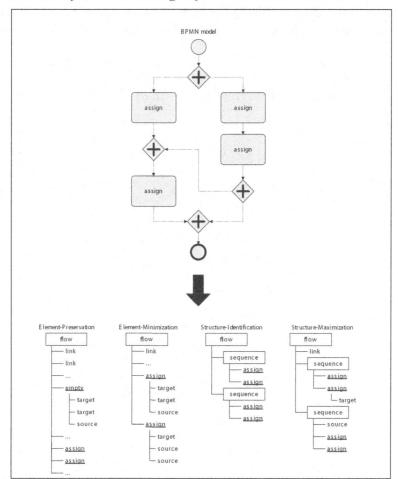

be applied for the transformation; if it is acyclic, the element-preservation strategy can be used. Figure 2 visualizes the different transformation strategies available.

Transformation Strategies from BPEL to BPMN

A transformation from BPEL to BPMN imposes restrictions only for one strategy. The flattening strategy can be utilized to transform any BPEL control flow to BPMN. BPEL structured activities are flattened to gateways and arcs without any nesting. It might be a problem that the nesting of structured activities gets lost, but in most cases, the resulting BPMN model graph is easy to understand.

The hierarchy-preservation strategy can be applied if the descriptive semantics of structured activities have to be preserved in the resulting BPMN model.

Figure 3. Transformation strategies from BPEL to BPMN

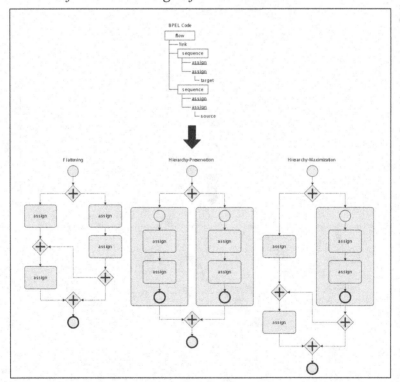

Each type of structured activity is mapped to a respective subprocess in BPMN. As a prerequisite for this strategy, the BPEL process may not include links.

The hierarchy-maximization strategy maps BPEL structured activities to subprocesses if no links exist in the model that would cross the boundaries of the subprocess. Accordingly, both the flattening and the hierarchy-preservation strategy have to be implemented.

Table 3 shows that arbitrary BPEL processes can be mapped to BPMN using the flattening or the hierarchy-maximization strategy. Since BPEL imposes more constraints on the process to be modeled than BPMN, a transformation is always feasible. Figure 3 visualizes the different transformation strategies available.

Further Approaches Toward Transformations from BPMN to BPEL

Beyond the transformation strategies introduced above, there are two further transformation approaches that can close the gap between BPEL and BPMN at least partially.

Ouyang, Dumas, et al. (2006) and Ouyang, van der Aalst, et al. (2006) propose a transformation of unstructured loops to event-condition-action rules that are implemented via BPEL event handlers. This approach yields a set of BPEL event handlers that a process calls on itself. This mechanism is able to capture unstructured loops. The structured part of the process graph can be encapsulated within BPEL event handlers. The unstructured part maps to messages sent from some place in the process to itself where it is forwarded to a corresponding event handler. A considerable benefit of this approach is that it abstracts from any potential topology of the BPMN model. On the other hand, the resulting BPEL code is difficult to comprehend and modify.

Another interesting idea for transformation is to derive a structured process model that is equivalent to the original, unstructured one. While such transformations have already been examined in the context of structured programming, they have only recently been discussed with respect to mapping unstructured process graphs to structured BPEL (Zhao, Hauser, Bhattacharya, Bryant, & Cao, 2006). However, even though unstructured process graphs that include only XOR splits and joins as rooting elements can always be transformed, this does not hold for arbitrary concurrency (Kiepuszewski, ter Hofstede,

& Bussler, 2000). Therefore, generating a structured model is not a general solution to the problem.

These two approaches represent rather new streams of research. It will be interesting to see how they can be combined with the existing set of transformation strategies in future work.

Conclusion

This chapter discussed conceptual mismatch between process modeling languages using the example of BPMN and BPEL. We used a generic approach incorporating various perspectives of the process model life cycle for identifying conceptual mismatch between process modeling languages employed in different stages of the BPM life cycle. In particular, our approach applies established evaluation theories and innovative transformation strategies in order to identify potential mapping issues in the form of the following:

- **Domain representation capability mismatch:** We showed how representation theory can be used to compare the representational capabilities of different process modeling languages in terms of divergences in the expressiveness of various aspects of domain semantics.
- **Control-flow support mismatch:** We showed how the workflow patterns framework can be used to identify discrepancies between process modeling languages in terms of their support for various aspects of control flow.
- **Process representation paradigm mismatch:** We showed how different representation paradigms underlying process modeling languages require different transformation strategies, and we sketched out the implications of the different strategies.

Referring back to our selected example, the analysis of the conceptual mismatch between BPMN and BPEL reveals that BPMN provides a much richer set of modeling constructs. A translation from technical BPEL to conceptual BPMN is therefore less a problem than in the opposite direction. Yet, BPMN is meant to be utilized as a visual notation for BPEL processes; however, as some of the BPMN constructs cannot be expressed in BPEL, a translation would imply a loss of information. For example, the missing BPEL support for a range of control-flow patterns that BPMN can support may, in a translation from BPMN to BPEL, lead to execution semantics that were not intended in the conceptual model. As a consequence, either process modeling in BPMN has to be restricted to those constructs that have an equivalent in BPEL, or a remodeling might be necessary on the level of BPEL in order to handle untranslatable constructs. In order to make the business process life cycle work, it seems to be a better option to restrict BPMN rather than to extend BPEL, as extensions of the latter may not be supported by existing standard-compliant work-flow engines.

Regarding directions to further research, we perceive this work to be a starting point on at least two counts. On the one hand, our approach may serve as a framework for a more detailed analysis of BPMN and BPEL (and other combinations of languages) using the theories and evaluation methods referred to in this chapter. In particular, we see a need to comparatively assess the varying domain representation capabilities, and control-flow support, of BPMN and BPEL in more detail, for example, by means of overlap analysis (Green et al., in press; Weber, 1997). On the other hand, our findings can serve as input to the formulation of more suitable transformation approaches that build upon the identified mismatches and are able to counteract the areas of concern our work was able to identify. Thereby, we foresee that the ultimate objective to provide a seamless integration of the various stages of the BPM life cycle by means of translatable process models can be supported.

In a more general sense, the conceptual mismatch between BPMN and BPEL reveals the lack of a general standardization strategy for business-process management that crosses different standardization bodies. Just recently, Nickerson and zur Muehlen (2006) carefully dissected the social process of standardization in various institutions related to Internet standards. Some of their key findings are that both devoted individuals and profit organizations seek to legitimize their concepts. Therefore, aesthetic values and commercial interests influence the standards adoption rather than macroeconomic considerations, let alone the potentially superior capabilities of one candidate over another. This makes not only the establishment of standards a time-consuming endeavor, but also the alignment of different standards affiliated in different standardization bodies. The case of BPMN and BPEL and the related standardization efforts indicates that a visionary standards road map would be valuable in particular for business-process modeling.

Acknowledgment

We gratefully acknowledge the fruitful contributions of our colleagues Michael Rosemann, Peter Green, Marta Indulska, Chris Manning, Petia Wohed, Wil van der Aalst, Arthur ter Hofstede, and Marlon Dumas to the evaluations of BPMN and BPEL by means of representation theory and work-flow patterns. Furthermore, we would like to thank Kristian Bisgaard Lassen and Uwe Zdun for the joint effort toward the identification of transformation strategies.

References

Andrews, T., Curbera, F., Dholakia, H., Goland, Y., Klein, J., Leymann, F., et al. (2003). *Business process execution language for Web services: Version 1.1.* Retrieved February 10, 2006, from http://xml.coverpages. org/BPELv11-May052003Final.pdf

Bider, I., & Johannesson, P. (2002). Modeling dynamics of business processes: Key for building next generation of business information systems. In S. Spaccapietra, S. T. March, & Y. Kambayashi (Eds.), *Conceptual modeling: ER 2002* (Vol. 2503, pp. 7-9). Tampere, Finland: Springer.

Bodart, F., Patel, A., Sim, M., & Weber, R. (2001). Should optional properties be used in conceptual modelling? A theory and three empirical tests. *Information Systems Research, 12*(4), 384-405.

BPMI.org. (2004). *Business process modeling notation (BPMN): Version 1.0. May 3, 20.* Retrieved March 2, 2005, from http://www.bpmn.org/

BPMI.org. (2005). *BPMN implementors and quotes.* Retrieved February 24, 2006, from http://www.bpmn.org/BPMN_Supporters.htm

BPMI.org & Object Management Group (OMG). (2006). *Business process modeling notation specification: Final adopted specification.* Retrieved February 20, 2006, from http://www.bpmn.org

Bunge, M. A. (2003). *Philosophical dictionary.* New York: Prometheus Books.

Curtis, B., Kellner, M. I., & Over, J. (1992). Process modeling. *Communications of the ACM, 35*(9), 75-90.

Davenport, T. H., & Short, J. E. (1990). The new industrial engineering: Information technology and business process redesign. *Sloan Management Review, 31*(4), 11-27.

Dehnert, J., & van der Aalst, W. M. P. (2004). Bridging the gap between business models and workflow specifications. *International Journal of Cooperative Information Systems, 13*(3), 289-332.

Dreiling, A., Rosemann, M., & van der Aalst, W. M. P. (2005). From conceptual process models to running workflows: A holistic approach for the configuration of enterprise systems. *2005 Pacific Asia Conference on Information Systems*, 363-376.

Dumas, M., van der Aalst, W. M. P., & ter Hofstede, A. H. M. (Eds.). (2005). *Process aware information systems: Bridging people and software through process technology.* Hoboken, NJ: John Wiley & Sons.

Ellison, M., & McGrath, G. M. (1998). Recording and analysing business processes: An activity theory based approach. *Australian Computer Journal, 30*(4), 146-152.

Fischer, L. (Ed.). (2005). *Workflow handbook 2005.* Lighthouse Point, FL: Future Strategies Inc.

Gao, Y. (2006). *BPMN-BPEL transformation and round trip engineering.* Retrieved June 30, 2006, from http://www.eclarus.com/pdf/BPMN_BPEL_Mapping.pdf

Gemino, A., & Wand, Y. (2005). Complexity and clarity in conceptual modeling: Comparison of mandatory and optional properties. *Data & Knowledge Engineering, 55*(3), 301-326.

Green, P., & Rosemann, M. (2000). Integrated process modeling: An ontological evaluation. *Information Systems, 25*(2), 73-87.

Green, P., & Rosemann, M. (2004). Applying ontologies to business and systems modeling techniques and perspectives: Lessons learned. *Journal of Database Management, 15*(2), 105-117.

Green, P., Rosemann, M., Indulska, M., & Manning, C. (in press). Candidate interoperability standards: An ontological overlap analysis. *Data & Knowledge Engineering.*

Gruber, T. R. (1993). A translation approach to portable ontology specifications. *Knowledge Acquisition, 5*(2), 199-220.

Guizzardi, G. (2005). *Ontological foundations for structural conceptual models* (Vol. 015). Enschede, the Netherlands: Telematica Instituut.

Hammer, M., & Champy, J. (1993). *Reengineering the corporation: A manifesto for business revolution.* New York: Harpercollins.

Jablonski, S., & Bussler, C. (1996). *Workflow management: Modeling concepts, architecture, and implementation.* London: Thomson Computer Press.

Katzenstein, G., & Lerch, F. J. (2000). Beneath the surface of organizational processes: A social representation. Framework for business process redesign. *ACM Transactions on Information Systems, 18*(4), 383-422.

Keller, G., Nüttgens, M., & Scheer, A.-W. (1992). *Semantische prozessmodellierung auf der grundlage "ereignisgesteuerter prozessketten (EPK)"* (Working Paper No. 89). Saarbrücken, Germany: Institut für Wirtschaftsinformatik, Universität Saarbrücken.

Kiepuszewski, B., ter Hofstede, A. H. M., & Bussler, C. (2000). On structured workflow modelling. In B. Wangler & L. Bergmann (Eds.), *Advanced information systems engineering: CAiSE 2000* (Vol. 1789, pp. 431-445). Stockholm, Sweden: Springer.

Kiepuszewski, B., ter Hofstede, A. H. M., & van der Aalst, W. M. P. (2003). Fundamentals of control flow in workflows. *Acta Informatica, 39*(3), 143-209.

Kloppmann, M., Koenig, D., Leymann, F., Pfau, G., Rickayzen, A., von Riegen, C., et al. (2005). *WS-BPEL extension for sub-processes: BPEL-SPE.* Retrieved August 14, 2006, from https://www.sdn.sap.com/irj/servlet/prt/portal/prtroot/docs/library/

Kung, C. H., & Sølvberg, A. (1986). Activity modeling and behavior modeling of information systems. In T. W. Olle, H. G. Sol, & A. A. Verrijn-Stuart (Eds.), *Information systems design methodologies: Improving the practice* (pp. 145-171). Amsterdam: North-Holland.

Mantell, K. (2005). *From UML to BPEL: Model driven architecture in a Web services world.* Retrieved June 30, 2006, from http://www-128.ibm.com/developerworks/webservices/library/ws-uml2bpel/

Mendling, J., Lassen, K. B., & Zdun, U. (2006). Transformation strategies between block-oriented and graph-oriented process modelling languages. In F. Lehner, H. Nösekabel, & P. Kleinschmidt (Eds.), *Multikonferenz wirtschaftsinformatik 2006, Band 2* (pp. 297-312). Berlin, Germany: GITO-Verlag.

Miers, D. (2003). The split personality of BPM. *The BPMG Newsletter, 11*(11), 1-22.

Milton, S., & Kazmierczak, E. (2004). An ontology of data modelling languages: A study using a common-sense realistic ontology. *Journal of Database Management, 15*(2), 19-38.

Mylopoulos, J. (1992). Conceptual modelling and telos. In P. Loucopoulos & R. Zicari (Eds.), *Conceptual modelling, databases, and CASE: An integrated view of information system development* (pp. 49-68). New York: John Wiley & Sons.

Nickerson, J. V., & zur Muehlen, M. (2006). The ecology of standards processes: Insights from Internet standard making. *MIS Quarterly, 30*(3), 467-488.

Nysetvold, A. G., & Krogstie, J. (2005). Assessing business process modeling languages using a generic quality framework. *CAiSE'05 Workshops, 1*, 545-556. Porto, Portugal: FEUP.

Ouyang, C., Dumas, M., Breutel, S., & ter Hofstede, A. H. M. (2006). Translating standard process models to BPEL. In E. Dubois & K. Pohl (Eds.), *Advanced information systems engineering: CAiSE 2006* (Vol. 4001, pp. 417-432). Luxembourg, Grand-Duchy of Luxembourg: Springer.

Ouyang, C., van der Aalst, W. M. P., Dumas, M., & ter Hofstede, A. H. M. (2006). From BPMN process models to BPEL Web services. *4th International Conference on Web Services*.

Petri, C. A. (1962). Fundamentals of a theory of asynchronous information flow. In C. M. Popplewell (Ed.), *IFIP Congress 62: Information Processing* (pp. 386-390). Munich, Germany: North-Holland.

Recker, J., Indulska, M., Rosemann, M., & Green, P. (2005). Do process modelling techniques get better? A comparative ontological analysis of BPMN. In *Proceedings of the 16ᵗʰ Australasian Conference on Information Systems*.

Recker, J., Indulska, M., Rosemann, M., & Green, P. (2006). How good is BPMN really? Insights from theory and practice. In *Proceedings of the 14ᵗʰ European Conference on Information Systems*.

Rosemann, M., & Green, P. (2002). Developing a meta model for the Bunge-Wand-Weber ontological constructs. *Information Systems, 27*(2), 75-91.

Rosemann, M., Recker, J., Indulska, M., & Green, P. (2006). A study of the evolution of the representational capabilities of process modeling grammars. In E. Dubois & K. Pohl (Eds.), *Advanced information sys-*

tems engineering: CAiSE 2006 (Vol. 4001, pp. 447-461). Luxembourg, Grand-Duchy of Luxembourg: Springer.

Russell, N., ter Hofstede, A. H. M., Edmond, D., & van der Aalst, W. M. P. (2005). Workflow data patterns: Identification, representation and tool support. In L. M. L. Delcambre, C. Kop, H. C. Mayr, J. Mylopoulos, & Ó. Pastor (Eds.), *Conceptual modeling: ER 2005* (Vol. 3716, pp. 353-368). Klagenfurt, Austria: Springer.

Russell, N., van der Aalst, W. M. P., ter Hofstede, A. H. M., & Edmond, D. (2005). Workflow resource patterns: Identification, representation and tool support. In Ó. Pastor & J. Falcão e Cunha (Eds.), *Advanced information systems engineering: CAiSE 2005* (Vol. 3520, pp. 216-232). Porto, Portugal: Springer.

Siau, K. (2004). Informational and computational equivalence in comparing information modeling methods. *Journal of Database Management, 15*(1), 73-86.

Smith, H., & Fingar, P. (2003). *Business process management: The third wave.* Tampa, FL: Meghan-Kiffer Press.

Uschold, M., & Grüninger, M. (1996). Ontologies: Principles, methods and applications. *The Knowledge Engineering Review, 11*(2), 93-136.

Van der Aalst, W. M. P., Dumas, M., ter Hofstede, A. H. M., & Wohed, P. (2002). *Pattern-based analysis of BPML (and WSCI)* (Tech. Rep. No. FIT-TR-2002-05). Brisbane, Australia: Queensland University of Technology.

Van der Aalst, W. M. P., & ter Hofstede, A. H. M. (2005). YAWL: Yet another workflow language. *Information Systems, 30*(4), 245-275.

Van der Aalst, W. M. P., ter Hofstede, A. H. M., Kiepuszewski, B., & Barros, A. P. (2003). Workflow patterns. *Distributed and Parallel Databases, 14*(1), 5-51.

Wand, Y. (1996). Ontology as a foundation for meta-modelling and method engineering. *Information and Software Technology, 38*(4), 281-287.

Wand, Y., Monarchi, D. E., Parsons, J., & Woo, C. C. (1995). Theoretical foundations for conceptual modelling in information systems Ddvelopment. *Decision Support Systems, 15*(4), 285-304.

Wand, Y., & Weber, R. (1990). An ontological model of an information system. *IEEE Transactions on Software Engineering, 16*(11), 1282-1292.

Wand, Y., & Weber, R. (1993). On the ontological expressiveness of information systems analysis and design grammars. *Journal of Information Systems, 3*(4), 217-237.

Wand, Y., & Weber, R. (1995). On the deep structure of information systems. *Information Systems Journal, 5*(3), 203-223.

Weber, R. (1997). *Ontological foundations of information systems*. Melbourne, Australia: Coopers & Lybrand & the Accounting Association of Australia and New Zealand.

Weske, M., van der Aalst, W. M. P., & Verbeek, H. M. V. (2004). Advances in business process management. *Data & Knowledge Engineering, 50*(1), 1-8.

White, S. A. (2005). Using BPMN to model a BPEL process. *BPTrends, 3*, 1-18.

Wohed, P., van der Aalst, W. M. P., Dumas, M., & ter Hofstede, A. H. M. (2003). Analysis of Web services composition languages: The case of BPEL4WS. In I.-Y. Song, S. W. Liddle, T. W. Ling, & P. Scheuermann (Eds.), *Conceptual modeling: ER 2003* (Vol. 2813, pp. 200-215). Chicago: Springer.

Wohed, P., van der Aalst, W. M. P., Dumas, M., ter Hofstede, A. H. M., & Russell, N. (2006). On the suitability of BPMN for business process modelling. In *Proceedings of the 4th International Conference on Business Process Management*.

Yu, E. S. K., Mylopoulos, J., & Lespérance, Y. (1996). AI models for business-process reengineering. *IEEE Expert: Intelligent Systems and Their Applications, 11*(4), 16-23.

Zhao, W., Hauser, R., Bhattacharya, K., Bryant, B. R., & Cao, F. (2006). Compiling business processes: Untangling unstructured loops in irreducible flow graphs. *International Journal of Web and Grid Services, 2*(1), 68-91.

Zur Muehlen, M. (2004). *Workflow-based process controlling: Foundation, design and application of workflow-driven process information systems*. Berlin, Germany: Logos.

Zur Muehlen, M., & Rosemann, M. (2004). Multi-paradigm process management. In *Proceedings of the CAiSE'04 Workshops in Connection with the 16th Conference on Advanced Information Systems Engineering* (Vol. 2, pp. 169-175).

Chapter X

Theories and Models:
A Brief Look at Organizational Memory Management

Sree Nilakanta, Iowa State University, USA

L. L. Miller, Iowa State University, USA

Dan Zhu, Iowa State University, USA

Abstract

This chapter introduces theories and models used in organizational memory. As organizations continue to automate their business processes and collect explosive amounts of data, researchers in knowledge management need to confront new opportunities and new challenges. In this chapter, we provide a brief review of the literature in organizational memory management. Some of the core issues of organizational memory management include organizational context, retention structure, knowledge taxonomy and ontology, organizational learning, distributed cognition and communities of practice, and so forth. As new information technologies are available to the design and implementation of organizational memory, we further present a basic framework of theories and models, focusing on the technological components and their applications in organizational memory systems.

Introduction

Organizational memory, a crucial component of an organization's knowledge ecosystem, plays a critical role in the overall performance and competitiveness of a business venture (March & Simon, 1958; Mort, 2001; Watson, 1998; Zhang, Tian, & Qi, 2006). In order to realize a benefit or strategic advantage, however, this knowledge must be properly managed. Consequently, many organizations are using formal knowledge management practices to improve performance. Knowledge management is best described as a process in which information is transformed into actionable knowledge and made available to the user (Allee, 1997). Effective knowledge management enables businesses to avoid repeating prior mistakes, to ensure the continued use of best practices, and to draw on the collective wisdom of its employees, past and present. Organizational memory is the collection of historical corporate knowledge that is employed for current use through appropriate methods of gathering, organizing, refining, and disseminating the stored information and knowledge (Ackerman & Halverson, 2000; Nevo & Wand, 2005).

The objectives of this chapter are to survey the organizational memory literature and present a basic framework on organizational memory systems (OMSs) and applications while focusing our attention on IT-based organizational memory. Research in organizational memory management deals with the creation, integration, maintenance, dissemination, and use of all kinds of knowledge within an organization (Alavi & Leidner, 1999; Cross & Baird, 2000). It is also confronted with new challenges because recent developments in information processing technologies have enhanced our ability to build the next generation of organizational memory management systems. Through our research studies, we found that much of the organizational memory is ignored or lost in the corporate collaborative processes in spite of the existence of several enterprise collaboration management tools. The consequence is that employees spend too much time re-creating common elements from online and off-line meetings, calendars, and various project-related activities.

In the next section, we review the literature of organizational memory management. Then we present a basic framework of technological components and their applications. Next we discuss some important research issues and future trends, and then conclude the chapter.

Organizational Memory

Organizational memory has been described as corporate knowledge that represents prior experiences and is saved and shared by corporate users. It includes both stored records (e.g., corporate manuals, databases, filing systems, etc.) and tacit knowledge (e.g., experience, intuition, beliefs; Nonaka & Takeuchi, 1995), and encompasses technical, functional, and social aspects of the work, the worker, and the workplace (Argote, McEvily, & Ray, 2003; Choy, Kwan, & Leong, 1999; Lee, Kim, Kim, & Cho, 1999). Organizational memory may be used to support decision making in multiple tasks and multiple user environments, for example, in construction (Ozorhon, Dikmen, & Birgonaul, 2005), in new product development (Akgun, Lynn, & Byrne, 2006), in machine learning and scheduling (Padman & Zhu, 2006), and in pursuing radical innovations (Johnson & Dilts, 2006). Walsh and Ungson (1991) refer to organizational memory as stored information from an organization's history that can be brought to bear on present decisions. By their definition, organizational memory provides information that reduces transaction costs, contributes to effective and efficient decision making, and is a basis for power within organizations. Researchers and practitioners recognize organizational memory as an important factor in the success of an organization's operations and its responsiveness to the changes and challenges of its environment (Huber, 1991; Huber, Davenport, & King, 1998).

Information technologies contribute to enable automated organizational knowledge management systems in two ways: either by making recorded knowledge retrievable or by providing vehicles for knowledgeable workers to share information (Chen, Hsu, Orwig, Hoopes, & Nunamaker, 1994; Olivera, 2000; Zhao, 1998). Explicitly dispersing an organization's knowledge through a variety of retention facilities (e.g., network servers, distributed databases, intranets, etc.) can make the knowledge more accessible to its members. Stein and Zwass (1995) suggest IT strategies can be used to maintain an extensive record of processes (through what sequence of events?), rationale (why?), context (under what circumstances?), and outcomes (how well did it work?). The availability of advanced information technologies increases the communicating and decision-making options for potential users.

Sandoe, Croasdell, Courtney, Paradice, Brooks, and Olfman (1998) use Giddens' (1984) definition of organizational memory to distinguish between discursive, practical, and reflexive memory, and they treat IT-based organizational memory as discursive. They argue that although IT-based memory operates at a discursive level, IT makes the discursive process of remembering more efficient by reducing the costs and effort associated with the storage of and access to an organization's memory. IT changes the balancing point in the trade-off between efficiency and flexibility, permitting organizations to be relatively more efficient for a given level of flexibility. Another advantage of IT-based memory is the opportunity to provide a historical narrative (or rationale) for significant organizational events that would otherwise be remembered in nondiscursive form. Furthermore, IT-based memory allows an organization to act in a rational manner through the discursive access to its major historical events and transformations. Additionally, Nevo and Wand (2005) note that IT-based organizational memory systems must deal not only with the location and source of memory, but also the context in which it occurs and is applicable. Finally, an OMS must address the tacit nature of some of the knowledge and the fact that the knowledge is volatile and has a finite life.

Mandiwalla, Eulgem, Mould, and Rao (1998) define an OMS to include a database management system (DBMS) that can represent more than transactional data, and an application that runs on top of the DBMS. They further describe the generic requirements of an OMS to include different types of memory, including how to represent, capture, and use organizational memory. Nemati, Steiger, Iyer, and Herschel (2002) illustrate that a knowledge warehouse combines three abilities: (a) an ability to efficiently generate, store, retrieve, and, in general, manage explicit knowledge in various forms, (b) an ability to store, execute, and manage the analysis tasks and their supporting technologies with minimal interaction and cognitive requirements from the decision maker, and (c) an ability to update the knowledge warehouse via a feedback loop of validated analysis output. The knowledge warehouse architecture has six major components: (a) the data or knowledge acquisition module, (b) the two feedback loops, (c) the extraction, transformation, and loading module, (d) a knowledge warehouse (storage) module, (e) the analysis workbench, and (f) a communications manager or user-interface module.

Haseman and Nazareth (2005) use the term collective memory to represent organization memory. They show that by building capabilities to share meeting data, prior decisions, and external sources of data into the collective memory repository, group decisions are enhanced. A skilled facilitator helps with collecting, maintaining, and processing group decisions and outcomes managed through the VisionQuest commercial software. These decisions and other memory contents are weighted and ranked by the participants and used to arrive at a consensus. Standard Web-based documents and personal database software complement the VisionQuest system to provide access to the group memory.

Technological Components and Applications

Organizational memory management must systematically deal with the creation, integration, maintenance, dissemination, and use of all kinds of knowledge within an organization (Cross & Baird, 2000). Although the system described in Haseman and Nazareth (2005) performed adequately to track the progress of an iterative decision-making process, it is lacking in many respects. The decisions and memory contents are ranked and weighted, but their use is limited to the extent of reviewing and revising the ranks and

Figure 1. Organizational knowledge model

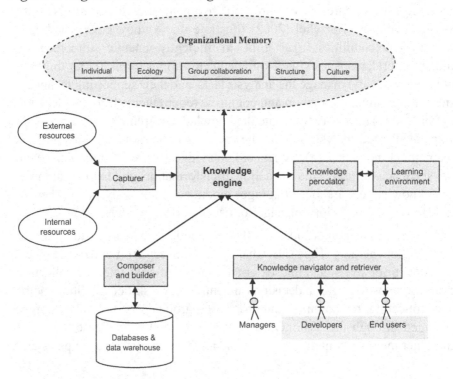

weights. Long-term use of such a system could result in massive amounts of data and there is no provision to aggregate or extract knowledge from the stored session details or decisions. Moreover, all users have to use a computer to enter their ratings and allocations, a limiting factor that we do not face in our model. In the absence of a computer, a user will have to maintain all of their allocations and ratings external to the system, which could result in loss of valuable information. For example, in most meetings, it is more likely that a human note taker is tasked with the recording of minutes, and he or she has at most access to a portable computer.

To bridge these issues, we propose a model that provides a more generic view of an organizational memory management system. Central to this model is a knowledge engine (KE) that works with the other components of the model to provide support for the creation and retrieval of knowledge. The capture component captures organizational memory information from internal and external sources. The composer and builder component facilitates the first-level composition or building of knowledge from the organization's various information collections. Without a retrieval and navigation system, any stored memory of knowledge would be useless. Key members of the organization, whether they are low-level users or executives, need a flexible yet comprehensible interface to the repository of organizational knowledge. In addition to these components, our model provides for the percolation of knowledge. It is built on the process of learning, either assisted through expert users or via automated machine-learning protocols. The individual components and the interaction of the key tasks of knowledge capture, composition, retrieval, and percolation offer a multitude of opportunities and issues.

Organizational memory is produced by a number of components, and captured and stored in various places. The capture of organizational memory is facilitated through a number of mechanisms such as meetings, e-mails, Web conferences, transaction processing, reporting systems, and so forth. The fine-grained information gets compiled and aggregated into relevant warehouses and knowledge bases through composer and builder systems and interfaces to the knowledge engine. The retriever and navigator systems and interfaces allow different types of users to access the stored organizational memory and knowledge. The percolator system and its interface enable users to extract and develop conclusions and hypotheses and build feedback loops for continuous learning. In addition to the interface between the knowledge engine and the four components, connection and continuity among the components also exist. The model creates a portal from the organization to its knowledge.

Specifically, the model automates the identification and distribution of relevant content, provides context sensitivity, and interacts intelligently with users, letting them profile, filter, and categorize information, and avails of the complex information infrastructure.

The proposed model is also designed to use work-group meetings as the primary data collection point. The assumption is that more traditional forms of data (databases, data warehouses, and report libraries) are easy to generate, and the major concern is to incorporate them in with the knowledge management process (Miller & Nilakanta, 1997). In most organizations, work-group meetings are central to the information-gathering and decision-making processes. The strength of the model lies in its ability to organize disparate information in a seamless fashion. Specifically, the model automates the identification and distribution of relevant content, provides content sensitivity, and interacts intelligently with users, letting them profile, filter, and categorize the complex information infrastructure.

Research Issues and Future Trends

Designing the ideal OMS is a difficult task, especially as definitions, technologies, and usage contexts continue to shift and evolve. A number of research issues need to be addressed.

- **Organizational context:** From an organizational context perspective, user communities and their work environments yield a number of issues. Focusing and reconciling group, interorganizational, and intra-organizational perspectives is necessary. For example, how will different types of users (individuals, groups, top management) perceive and use an OMS? Will organizational roles and power affect the use of an OMS? Another issue is the role of individual memories. Users may have their personal collections of memory that are both private and public. These raise a number of relevant questions as well. Where do individually held memories fit in the OMS? Are they redundant? How can they be used? What are the legal and social implications of storing and using them?

- **Retention structure:** According to Walsh and Ungson (1991), an OMS is composed of knowledge compiled from individuals, groups, organizational structures, ecology, and culture. Each of these requires appropriate capture, encoding, and integration mechanisms. What are the cost implications? How long will the information be kept? From a data source perspective, information sources can be internal or external to the firm. Also, the sources may be private or public. In addition, the value of information will be affected by its various quality attributes. Therefore, questions arise as to how different sources of information will be valued in an organization's memory. What data management policies will be required? Retaining organizational memory typically implies some type of storage device. In the foreseeable future, information storage will always involve costs associated with storage media, the time needed to access the selected media, and administrative costs of maintaining the information. Organizations will need tools that will help them evaluate the costs and benefits of storing all forms and types of memory. For example, 1 second of video at 24-bit color depth (30 frames) needs about 27MB of space. This means that about 3 hours of video could require a 10-Gigabyte medium with a 20:1 compression. As a result, even though storage requirements are expected to decline rapidly as newer compression algorithms and methods are developed, storage will always be an issue. Incorporating video data quickly tilts the balance away from comprehensiveness. Increasing comprehensiveness also increases the potential for information overload. Assuming limited storage space, who decides what information should be kept? What is the mechanism and criteria for filtering? How can bias be avoided?

- **Knowledge taxonomy and ontology:** Widely held assumptions about data imply that the more organizational memory we store, the harder it becomes to locate a specific memory item of interest. Therefore, organizational-memory conceptual models will need a retrieval and classification mechanism built around some form of domain ontology. Hwang and Salvendy (2005) used general and domain-specific ontology models to represent historical events (memories of events) and found that the ontology models help in organizational learning. Abel, Benayache, Lenne, Moulin, Barry, and Chaput (2004) also found domain-specific ontology models useful in e-learning tasks. This raises questions about the diversity of domains, and models of ontology that are applicable. Integration, aggregation, and reintegration also pose challenges. For example, if information about the same topic is stored in multiple formats, for example, in database and multimedia format, users will need tools to reintegrate or "re-understand" and synchronize the memory. Knowledge taxonomy is also useful in designing and developing suitable mechanisms for its management and use. OMS components can be expected to behave differently, for example, in dealing with tacit knowledge than with explicit knowledge. Alavi and Leidner (2001) presented a number of research questions related to the four areas of knowledge management, namely, knowledge creation, storage and retrieval, transfer, and application. These four areas correspond to the four core components of our OMS. Chou (2005) found that organizational-level changes have more effect on knowledge creation. Furthermore, the research showed that the ability to put the knowledge into practice is more important than the knowledge itself, thus reiterating the need to have adequate mechanisms for creating and retrieving knowledge. What mechanisms and best practices are relevant in knowledge creation and retrieval? Because of the inherent value embedded in an OMS, the information asset needs to be secured and controlled to protect its integrity and safeguard the privacy of its creators and users. Alarcon, Guerrero, and Pino (2005) proposed a four-level privacy model for using organizational memory. At the "no privacy" level, information is widely available for use, and collaboration becomes seamless. As the privacy level ratchets to fully restricted information, memory needs interpretation and qualitative assessments. The need to impose controls on the use and dissemination of memory raises issues related to privacy and security. What is the acceptable level of security and control? What privacy and security models are applicable? Finally, information and knowledge can become obsolete over time. Information life-cycle management is an approach firms have started to apply in this regard.

- **Organizational learning:** The core piece of the proposed model, the knowledge engine, focuses on the creation, storage and integration, retrieval, and repurposing of the assimilated knowledge. The set of tools and mechanisms rely on several knowledge management theories and assumptions. Both automatic learning and human-assisted learning are needed to maintain a growing collection of useful memories. While the major question an organizational memory model should address is whether the knowledge can improve organizational performance, several additional issues may also be raised concerning OMS design and implementation. Essentially, an OMS enables the capture, storage, and integration of knowledge and best practices so that these may be retrieved, analyzed, consumed, and repurposed by users. In order to establish appropriate design and use criteria, the OMS must correspond to well-grounded theories of knowledge elicitation and use. Cognitive science and transactive memory models are useful here (Zhu & Prietula, 2002). Transactive memory consists of the information stored in each individual member's memory and the awareness of the type of information held by other members of the group. The encoding, storage, and retrieval of transactive information are facilitated by communications and interactions among the group members.

- **Distributed cognition and communities of practice:** Ackerman and Halverson (2004) take a critical view of prior research on OM and argue for a theoretical base to properly define and empirically validate future research. They state that as sociotechnical systems, organizations and their memories conform to social structures and norms while employing technical models. They use the theory of distributed cognition to develop a theoretical foundation for organizational memory. The basic tenets of this theory are that knowledge evolves from a community of practice and that cognition and inferences result from the shared meaning among the participants (hence the distribution; Hollan, Hutchins, & Kirsch, 2000). Communities of practice fulfill a number of functions with respect to the creation, accumulation, and diffusion of knowledge in an organization through the exchange and interpretation of information, by retaining knowledge, by stewarding competencies, and by providing homes for identities (Wenger, 1998). Collective thinking creates knowledge that otherwise would not be evident. Additionally, changes in the state of the memory, as in changing from internal to external representations via artifact changes or through the movement of information among the participants (trajectory of information), are necessary to fully utilize an OM. A cycle of changes comprising contextualization to decontextualization and again to recontextualization of the information object takes place as organizational members relive their experience through the stored information object or artifact. An essential feature of knowledge management systems is this capability to change the state of the information object.

Conclusion

Technological changes and shifting demands make rapid learning essential in organizations. The advent and increasingly wide utilization of wide-area-network tools such as the Internet and World Wide Web provide access to greater and richer sources of information. Local area networks and intranets give organizations ways to store and access memory and knowledge that is specific to the organization. Used effectively, these tools support the concept of organizational memory.

Currently, there is a strong need for developing sound design and methodologies for the Net-enabled business. Any model is useful only insofar as it helps to answer relevant and valid questions. A number of research issues have been identified in this chapter. The discussion of these research questions calls for multidisciplinary approaches that integrate the technologies from a number of fields such as business, computer science, organization science, and cognitive psychology.

In an era of rapid and continuous change, our capacity to continue to shape the future will rely on our ability to learn, to create knowledge, and to adapt (Zhu, Prietula, & Hsu, 1997). We need to carefully study the organizational learning of business processes so as to deliver full value to an intelligent organization. To this end, researchers in organizational memory management must address the issues of knowledge management successfully.

Acknowledgment

This research is partially supported under summer research grants from Icube and Iowa State University.

References

Abel, M. H., Benayache, A., Lenne, D., Moulin, C., Barry, C., & Chaput, B. (2004). Ontology-based organizational memory for e-learning. *Educational Technology & Society, 7*(4), 98-111.

Ackerman, M., & Halverson, C. (2004). Organizational memory as objects, processes, and trajectories: An examination of organizational memory in use. *Computer Supported Cooperative Work (CSCW), 13*(2), 155-189.

Ackerman, M. S., & Halverson, C. A. (2000). Reexamining organizational memory. *Communications of the ACM, 43*(1), 58-64.

Akgun, A. E., Lynn, G. S., & Byrne, J. C. (2006). Antecedents and consequences of unlearning in new product development teams. *Journal of Product Innovation Management, 23*(1), 73-88.

Alarcon, R. A., Guerrero, L. A., & Pino, J. A. (2005). *Temporal blurring: A privacy model for OMS users.* Paper presented at User Modeling 2005.

Alavi, M., & Leidner, D. E. (1999). Knowledge management systems: Issues, challenges, and benefits. *Communications of the Association of Information Systems, 1*, 1-37.

Alavi, M., & Leidner, D. E. (2001). Review: Knowledge management and knowledge management systems. Conceptual foundations and research issues. *MIS Quarterly, 25*(1), 107-136.

Allee, V. (1997). *The knowledge evolution: Expanding organizational intelligence.* Butterworth-Heinemann.

Argote, L., McEvily, B., & Ray, R. (2003). Managing knowledge in organizations: An integrative framework and review of emerging themes. *Management Science, 49*(4), 571-583.

Chen, H., Hsu, P., Orwig, R., Hoopes, L., & Nunamaker, J. (1994). Automatic concept classification of text from electronic meetings. *Communications of the ACM, 37*(10), 56-73.

Chou, S. W. (2005). Knowledge creation: Absorptive capacity, organizational mechanisms, and knowledge storage/retrieval capabilities. *Journal of Information Science, 31*(6), 453-465.

Choy, M., Kwan, M.-P., & Leong, H. V. (1999). Distributed database design for mobile geographical applications. *Journal of Database Management, 11*(1), 3-17.

Cross, R., & Baird, L. (2002). Technology is not enough: Improving performance by building organizational memory. *Sloan Management Review, 41*(3), 69-78.

Haseman, W. D., & Nazareth, D. L. (2005). Implementation of a group decision support system utilizing collective memory. *Information and Management, 42*, 591-605.

Hollan, J., Hutchins, E., & Kirsh, D. (2000). Distributed cognition: Toward a new foundation for human-computer interaction research. *ACM Transactions on Computer-Human Interaction, 7*(2), 174-196.

Huber, G. (1991). Organizational learning: The contributing processes and literature. *Organization Science, 2*, 88-115.

Huber, G., Davenport, T. H., & King, D. R. (1998). *Perspectives on organizational memory.* Paper presented at the 31st Annual Hawaii International Conference on System Sciences Task Force on Organizational Memory, HI.

Hwang, S.-Y., & Yang, W.-S. (2002). On the discovery of process models from their instances. *Decision Support Systems, 34*(1), 41.

Johnson, J. H., & Dilts, D. M. (2006). *Acquire and forget: The conflict of information acquisition and organizational memory in the development of radical innovations.* Paper presented to the American Marketing Association.

Lee, H., Kim, J., Kim, Y. G., & Cho, S. H. (1999). A view-based hypermedia design methodology. *Journal of Database Management, 10*(2), 3-13.

Mandiwalla, M., Eulgem, S., Mould, C., & Rao, S. V. (1998). Organizational memory system design. *Proceedings of the Thirty-First Annual Hawaii International Conference on System Sciences.*

March, J. G., & Simon, H. A. (1958). *Organizations.* New York.

Markus, M. L. (2001). Toward a theory of knowledge reuse: Types of knowledge reuse situations and factors in reuse success. *Journal of Management Information System, 18*(1), 57-94.

Miller, L. L., & Nilakanta, S. (1997). Tools for organizational decision support: The design and development of an organizational memory system. In *Proceedings of the Thirtieth Annual Hawaii International Conference on System Sciences* (pp. 360-368).

Mort, J. (2001). Nature, value, and pursuit of reliable corporate knowledge. *Journal of Knowledge Management, 5*(3), 222-230.

Nemati, H. R., Steiger, D. M., Iyer, L. S., & Herschel, R. T. (2002). Knowledge warehouse: An architectural integration of knowledge management, decision support, artificial intelligence and data warehousing. *Decision Support Systems, 33*(2), 143-161.

Nevo, D., & Wand, Y. (2005). Organizational memory information systems: A transactive memory approach. *Decision Support Systems, 39*, 549-562.

Nonaka, I., & Takeuchi, H. (1995). *The knowledge creating company: How Japanese companies create the dynamics of innovation.* New York: Oxford University Press.

Olivera, F. (2000). Memory systems in organizations: An empirical investigation of mechanisms for knowledge collection, storage and access. *The Journal of Management Studies, 37*(6), 811-830.

Ozorhon, B., Dikmen, I., & Birgonaul, M. T. (2005). Organizational memory formation and its use in construction. *Building Research and Information, 33*(1), 67-79.

Padman, R., & Zhu, D. (2006). Knowledge integration using problem spaces: A study in resource-constrained project scheduling. *Journal of Scheduling, 9*(2), 133-152.

Sandoe, K., Croasdell, D. T., Courtney, J., Paradice, D., Brooks, J., & Olfman, L. (1998). Additional perspectives on organizational memory. In *Proceedings of the Thirty-First Annual Hawaii International Conference on System Sciences Task Force on Organizational Memory.*

Stein, E. (1995). Organizational memory: Review of concepts and recommendations for management. *International Journal of Information Management, 15*(2), 17-32.

Stein, E. W., & Zwass, V. (1995). Actualizing organizational memory with information systems. *Information Systems Research, 6*(2), 85-117.

Walsh, J. P., & Ungson, G. R. (1991). Organizational memory. *The Academy of Management Review, 16*(1), 57-91.

Watson, R. T. (1998). *Data management, databases and organizations* (2nd ed.). New York.

Wenger, E. (1998). *Communities of practice: Learning, meaning, and identity.* New York: Cambridge.

Zhang, L., Tian, Y., & Qi, Z. (2006). Impact of organizational memory on organizational performance: An empirical study. *The Business Review, 5*(1), 227.

Zhao, J. L. (1998, August). Knowledge management and organizational learning in workflow systems. In *Proceedings of the AIS Americas Conference on Information Systems.*

Zhu, D., & Prietula, M. J. (2002). Intelligent architectures for knowledge sharing: A Soar example and general issues. In *Proceedings of FLAIRS Conference* (pp. 318-320).

Zhu, D., Prietula, M., & Hsu, W. (1997). When processes learn: Steps toward crafting an intelligent organization. *Information Systems Research, 8*(3), 302-317.

About the Contributors

Keng Siau is the E.J. Faulkner professor of MIS at UNL. He is currently serving as the editor-in-chief of the *Journal of Database Management* and as the director of the UNL-IBM program. He received his PhD degree from the University of British Columbia (UBC), where he majored in MIS and minored in cognitive psychology. His master's and bachelor's degrees are in computer and information sciences from the National University of Singapore. Dr. Siau has over 200 academic publications. He has published more than 90 refereed journal articles, and these articles have appeared (or are forthcoming) in journals such as *Management Information Systems Quarterly*; *Communications of the ACM*; *IEEE Computer*; *Information Systems*; *ACM SIGMIS's Data Base*; *IEEE Transactions on Systems, Man, and Cybernetics*; *IEEE Transactions on Professional Communication*; *IEEE Transactions on Information Technology in Biomedicine*; *IEICE Transactions on Information and Systems*; *Data and Knowledge Engineering*; *Decision Support Systems*; *Journal of Information Technology*; *International Journal of Human-Computer Studies*; *International Journal of Human-Computer Interaction*; *Behaviour and Information Technology*; *Quarterly Journal of Electronic Commerce*; and others. In addition, he has published more than 100 refereed conference papers (including 10 ICIS papers), edited or co-edited more than 15 scholarly and research-oriented books, edited or coedited nine proceedings, and written more than 20 scholarly book chapters. He served as the organizing and program chair of the International Workshop on Evaluation of Modeling Methods in Systems Analysis and Design (EMMSAD, 1996-2005). He also served on the organizing committees of AMCIS 2005, ER 2006, and AMCIS 2007. For more information on Dr. Siau, please visit his Web site at http://www.ait.unl.edu/siau/.

* * * * *

Mehmet N. Aydin is an assistant professor at the Department of Information Systems and Change Management at the Faculty of Business, Public Administration, and Technology, University of Twente, The Netherlands. He holds a PhD from the University of Twente where he has been teaching several courses about business process support, electronic commerce, and information systems development (ISD) methodologies. Before joining the university, he worked for Accenture with the Communication and Hi-Tech Service Line. His research interests include agile information systems development, the foundation and modeling of business services, and method engineering. He is involved in consultancy concerning the design of ISD methods in various organizations in financial, government, and hi-tech industries. In 2006 he served as an international visiting scholar at Ryerson University, Toronto, Ontario (Canada). His works appear as book chapters, articles in several journals, and in IFIP and AMCIS proceedings.

Jian Cai is an assistant professor of management information systems (MIS) at the Guanghua School of Management at Peking University (China). His primary areas of research include IT strategy, knowledge management, and business performance management. He has published in various academic journals and authored three books. Professor Cai earned a BE in manufacturing from Tsinghua University, an MS in computer engineering, and a PhD in intelligent design systems from the University of Southern California.

John Erickson is an assistant professor in the College of Business Administration at the University of Nebraska – Omaha (USA). His current research interests include the study of UML as an OO systems development tool, software engineering, and the impact of structural complexity upon the people and systems involved in the application development process.

He has published in *Communications of the ACM*, the *Journal of Database Management*, and several refereed conferences such as AMCIS, ICIS WITS, EMMSAD, and CAiSE. Erickson has also authored materials for a distance education course at the University of Nebraska, Lincoln (UNL), collaborated on a book chapter, and co-chaired minitracks at several AMCIS conferences. He has served as a member of the program committee for EMMSAD and is on the editorial review board for the *Journal of Database Management* and the *Decision Sciences Journal*.

Terry Halpin (BSc, DipEd, BA, MLitStud, PhD) is a distinguished professor and vice president (conceptual modeling) at Neumont University (USA). After many years in academia, he worked on data modeling technology at Asymetrix Corporation, InfoModelers Inc., Visio Corporation, and Microsoft Corporation before returning to academia to develop data models and curricula to facilitate application development using a business-rules approach to informatics. His research focuses on conceptual modeling and conceptual query technology. His doctoral thesis formalized object-role modeling (ORM/NIAM). He has authored over 130 technical publications and five books, including *Information Modeling and Relational Databases* and *Database Modeling with Microsoft Visio for Enterprise Architects*, and has coedited three books on research issues in information systems modeling. He is a member of IFIP WG 8.1 (information systems) and several academic program committees, is an editor or reviewer for several academic journals, and has presented seminars and tutorials at dozens of international conferences.

Frank Harmsen is a principal consultant with Capgemini IT Performance Consulting (USA), an affiliated researcher at the University of Utrecht, and a guest lecturer at the University of Twente. He is involved in research and consultancy concerning the improvement of IT processes and IT organizations, including situational method engineering, IT governance, and organizational change management. He holds an MSc in computer science and business administration from Radboud University of Nijmegen and a PhD in computer science from the University of Twente. In 1996, he worked as a visiting researcher for the Tokyo Institute of Technology. Dr. Harmsen has published around 20 papers on situational method engineering and has served on the program committee of several conferences.

Stijn Hoppenbrouwers received master's degrees in English (1993, Utrecht, The Netherlands) and linguistics (1994, Bangor, Wales). In December 2003, he obtained his PhD degree in computer science at Nijmegen. He now works as an assistant professor at the Nijmegen Institute for Computing and Information Sciences at the Radboud University Nijmegen, The Netherlands. His main focus is on processes for modeling in the context of system development. He teaches various topics, including requirements engineering and quality of information systems.

Kalle Lyytinen is the Iris S. Wolstein professor at the Weatherhead School of Management at Case Western Reserve University (USA) and an adjunct professor at the University of Jyväskylä. He is also the editor in chief of the *Journal of AIS*. Kalle was educated at the University of Jyväskylä, Finland, where he has studied computer science, accounting, statistics, economics, theoretical philosophy, and political theory. He has a bachelor's degree in computer science and a master's and PhD in economics (computer science). He has published eight books, over 50 journal articles, and over 80 conference presentations and book chapters. He is well known for his research in computer-supported system design and modeling, system failures and risk assessment, computer-supported cooperative work, and the diffusion of complex technologies. He is currently researching the development and management of digital services and the evolution of virtual communities. Prior to joining Weatherhead, Kalle was the dean of the Faculty of Technology at the University of Jyväskylä. He has held visiting positions at the Royal Technical Institute of Sweden, the London School of Economics, the Copenhagen Business School in Denmark, Hong Kong University of Science and Technology, Georgia State University, Aalborg University, the University of Pretoria (South Africa), and Erasmus University in The Netherlands.

Jan Mendling (1976) is a PhD student at the Institute of Information Systems and New Media at the Vienna University of Economics and Business Administration, Austria. His research interests include business process management, enterprise modeling, and work-flow standardization. He is coauthor of the EPC markup language (EPML) and co-organizer of the XML4BPM (Extensible Markup Language for Business Process Management) workshop series.

Hilkka Merisalo-Rantanen has an MSc in economy. She is a research fellow at the Graduate School of Electronic Business and Software Industry, and is a PhD candidate in information systems science at the Helsinki School of Economics, Finland. Her research interests include information systems development methods, stakeholder and end-user participation and collaboration throughout the information system life cycle, and multicustomer-multivendor information system development projects. She has worked over 20 years on various tasks of information systems development, consultancy, and project management in leading Finnish companies. She has published in the *Journal of Database Management*, *IEEE Transactions on Professional Communication*, and *Group Decision and Negotiation* as well as in conference proceedings (GDN, IRIS).

L. L. Miller received a BA (1967) and an MA (1974) in mathematics at the University of South Dakota, and a PhD (1980) in computer science from Southern Methodist University. At Iowa State University (USA), he was an assistant professor (1984-1987), an associate professor (1987-1991), and a professor (1991-present) in computer science. He served as department chairman of computer science from 1998 to 2001. His major research interests are in object-oriented databases, organizational decision-support systems, data warehouses, database semantics, organizational memory, parallel searching methods, multiagent systems, database design, data mining, and computational biology. Dr. Miller is currently looking at the developing infrastructure for providing geospatial data to field-survey and exploration applications. His other work on geospatial data focuses on developing accuracy models. His current activity in organizational decision-support systems centers on the use of object-based database systems to support the decision process. Dr. Miller's work on organizational memory is focused on the capture of organizational semantics and the integration of corporate documents into the meeting process.

Isabelle Mirbel received a PhD degree in computer science from the University of Nice-Sophia Antipolis in 1996. She is an assistant professor of computer engineering at the University of Nice-Sophia Antipolis. She is a member of the I3S Laboratory (UMR 6070, CNRS-UNSA). Her research interests include information system modeling and integration, work-flow design, and situational method engineering. She has published several papers in international journals and conferences, and contributed to several books.

Sree Nilakanta is an associate professor of management information systems at Iowa State University (USA). He received his MBA and PhD in information systems from the University of Houston. Dr. Nilakanta also holds a BS in mechanical engineering from Madras University. Dr. Nilakanta's research straddles both behavioral and technical domains of information systems. His primary research interests are in technology innovation, database management, and organizational memory. His research has appeared in *Management Science*, *Journal of Management Information Systems*, *Decision Support Systems*, *Information & Management*, *Journal of Software and Information Technology*, *Journal of Strategic Information Systems*, *Omega*, and others.

Erik Proper is a professor at the University of Nijmegen (The Netherlands). His main research interests include system theory, system architecture, business and IT alignment, conceptual modeling, information retrieval, and information discovery. Erik received his master's degree from the University of Nijmegen in May 1990, and his PhD (with distinction) from the same university in April 1994. His teaching includes courses on information architecture and the modeling of organizations.

Jan Recker (1979) is a PhD student with the business process management research group of the Faculty of Information Technology at Queensland University of Technology, Brisbane (Australia). His research interests include business process modeling, conceptual model evaluation, process configuration, and reference modeling for enterprise systems. He has published more than 20 refereed journal papers, book chapters, and conference papers on these topics.

Iris Reinhartz-Berger is a faculty member at the Department of Management Information Systems, University of Haifa (Israel). She received her PhD in information management engineering from the Technion, Israel Institute of Technology. Her research interests include conceptual modeling, modeling languages and techniques for analysis and design, domain analysis, and development processes. Her work has been published in journals and international conference proceedings.

Matti Rossi is an acting professor of information systems and director of the electronic business program for professionals (Muuntokoulutus) at Helsinki School of Economics (Finland). He has worked as a research fellow at Erasmus University Rotterdam and as a visiting assistant professor at Georgia State University, Atlanta. He received his PhD degree in business administration from the University of Jyväskylä in 1998. He has been the principal investigator in several major research projects funded by the Technological Development Center of Finland and the Academy of Finland. His research papers have appeared in journals such as CACM, the *Journal of AIS*, *Information and Management*, and *Information Systems*, and over 30 of them have appeared in conference proceedings such as ICIS, HICSS, and CAiSE. More information is located at http://hkkk.fi/~mrossi/.

Pnina Soffer is a faculty member of the MIS Department at the University of Haifa (Israel). She received her PhD from the Technion, Israel Institute of Technology in 2002 developing a requirement-driven approach to the alignment of enterprise processes and an ERP (enterprise resource planning) system. Soffer has industrial experience as a production engineer and as an ERP consultant. Her current research areas are business process modeling, conceptual modeling, and requirements engineering.

Robert A. Stegwee is professor of e-health architecture and standards at the Faculty of Business, Public Administration, and Technology of the University of Twente (The Netherlands) and a principal consultant with Capgemini Health Services, The Netherlands. He was the former head of the Department of Business Information Systems at the University of Twente. He holds an MSc in computer science with a specialization in management information systems (cum laude, with honors) from the University of Amsterdam and a doctorate in organization and management from the University of Groningen. He is a member of the board of HL7, The Netherlands. His consultancy experience includes architecture for (regional) health information systems, decision-support and knowledge systems, process analysis and redesign, and the development of management information. He is active in editing international journals and has published many articles.

Arnon Sturm is a faculty member at Ben-Gurion University of the Negev (Israel). His research focuses on software engineering issues, in particular, conceptual modeling and development processes. During the last years his major research area has been domain engineering. Arnon has also gained extensive experience in developing software systems in industry and served as a member of software engineering groups that deal with system development problems.

Tuure Tuunanen received his doctoral degree in information systems science at the Helsinki School of Economics (Finland) in 2005. His current research interests lie in the area of IS development methods and processes, requirements engineering, and the convergence of IS and marketing disciplines in design. He is currently a senior lecturer at The University of Auckland Business School. His research has been published in *Information & Management*, *Journal of Database Management*, *Journal of Information Technology Theory and Application*, and *Journal of Management Information Systems*. In addition, his work has appeared in a variety of conference proceedings within his research interest areas, such as eCOMO, DESRIS, ISD, HICSS, Mobility Roundtable, RE, WeB, and WITS. Up-to-date information about his research is available at http://www.tuunanen.fi.

Patrick van Bommel received his master's degree in computer science in 1990 and his PhD in 1995 from the Faculty of Mathematics and Computer Science at the University of Nijmegen (The Netherlands). He is currently an assistant professor at the same university. He teaches courses on the foundations of databases and information systems, and on information analysis and design, and also supervises a university-based semicommercial student software house. His main research interests include information modeling and information retrieval.

Theo van der Weide received his master's degree at the Technical University Eindhoven (The Netherlands), in 1975, and his PhD in mathematics and physics from the University of Leiden (The Netherlands), in 1980. He is currently a professor in the Nijmegen Institute for Computing and Information Sciences at the Radboud University Nijmegen (The Netherlands), and head of the Department of Information and Knowledge Systems. His main research interests include information systems, information retrieval, hypertext, and knowledge-based systems.

Jos van Hillegersberg is a professor at the Faculty of Business, Public Administration and Technology, University of Twente (The Netherlands). His research interests include software development for e-business (CBD, EAI, UML, software process improvement), global software development, ICT support for the coordination of global teams, and ICT architectures. He worked earlier at the IBM Knowledge Based Center, as a visiting researcher at the CIS Department of Georgia State University, Atlanta (USA), as a visiting professor at Florida International University, and at AEGON Bank on the development of an e-banking system. Professor Hillegersberg is currently the head of the Department of Information Systems and Change Management and holds the chair in Design and Implementation of Information Systems at the University of Twente. He is active in editing international journals and has published many articles in journals including *Communications of the ACM, Journal of Information Technology*, and *Journal of Product Innovation Management*.

Dan Zhu is an associate professor at the Iowa State University (USA). She obtained her PhD degree from Carnegie Mellon University. Dr. Zhu's research has been published in the *Proceedings of National Academy of Sciences, Information System Research, Naval Research Logistics, Annals of Statistics, Annals of Operations Research*, and others. Her current research focuses on developing and applying intelligent and learning technologies to business and management.

Index